C Programming for Arduino

Learn how to program and use Arduino boards
with a series of engaging examples, illustrating
each core concept

Julien Bayle

PUBLISHING

BIRMINGHAM - MUMBAI

C Programming for Arduino

First published: May 2013

Production Reference: 1070513

Published by Packt Publishing Ltd.
Livery Place
35 Livery Street
Birmingham B3 2PB, UK.

ISBN 978-1-84951-758-4

www.packtpub.com

Cover Image by Asher Wishkerman (a.wishkerman@mpic.de)

Credits

Author
Julien Bayle

Reviewers
Darwin Grosse
Pradumn Joshi
Phillip Mayhew
Glenn D. Reuther
Steve Spence

Acquisition Editor
Edward Gordon
Erol Staveley

Lead Technical Editor
Susmita Panda

Technical Editors
Worrell Lewis
Varun Pius Rodrigues
Lubna Shaikh
Sharvari Baet

Copy Editors
Laxmi Subramanian
Sajeev Raghavan
Insiya Morbiwala
Brandt D'mello
Aditya Nair
Alfida Paiva

Project Coordinator
Leena Purkait

Proofreaders
Claire Cresswell-Lane
Martin Diver

Indexer
Tejal R. Soni

Graphics
Ronak Dhruv

Production Coordinator
Pooja Chiplunkar

Cover Work
Pooja Chiplunkar

About the Author

Julien Bayle completed his Master's degree in Biology and Computer Sciences in 2000. After several years working with pure IT system design, he founded Design the Media in early 2010 in order to provide his own courses, training, and tools for art fields. As a digital artist, he has designed some huge new media art installations, such as the permanent exhibition of La Maison des Cinématographies de la Méditerranée (Château de la Buzine) in Marseille, France, in 2011. He has also worked as a new media technology consultant for some private and public entities. As a live AV performer, he plays his cold electronic music right from New York to Marseille where he actually lives. The Arduino framework is one of his first electronic hardware studies since early 2005, and he also designed the famous protodeck controller with various open source frameworks. As an Art and Technology teacher also certified by Ableton in 2010, he teaches a lot of courses related to the digital audio workstation Ableton Live, the real-time graphical programming framework Max 6, and Processing and Arduino.

As a minimalist digital artist, he works at the crossroads between sound, visual, and data. He explores the relationship between sounds and visuals through his immersive AV installations, his live performances, and his released music. His work, often described as "complex, intriguing, and relevant", tries to break classical codes to bring his audience a new vision of our world through his pure digital and real-time-generated stimuli.

He's deeply involved in the open source community and loves to share and provide workshops and masterclasses online and on-site too. His personal website is http://julienbayle.net.

Acknowledgement

I would like to thank my sweet wife Angela and our daughter Alice for having been my unconditional supporters. Special thanks to our son Max, who was born between the writing of Chapter 11 and Chapter 12!

I would also like to thank my two great friends Laurent Boghossian and Denis Laffont because they were there for me all through the course of this huge project with their advices, jokes, and unconditional support.

I would like to extend many thanks to two very nice persons and friends whom I asked to review this book for me: Glenn D. Reuther and Darwin Grosse.

I thank the following great programmers who coded some libraries that have been used in this book: Marcello Romani (the SimpleTimer library), Juan Hernandez (the ShiftOutX library), Thomas Ouellet Fredericks (the Bounce library), Tim Barrass (the Mozzi library), David A. Mellis from MIT (the PCM library), Michael Margolis and Bill Perry (the glcd-arduino library), and Markku Rossi (Arduino Twitter Library with OAuth Support).

I want to thank the creators of the following powerful frameworks used in this book besides the Arduino framework itself: Max 6, Processing, and Fritzing.

Lastly, I'd like to hug Massimo Banzi and Arduino's project team for having initiated this great project and inspired us so much.

About the Reviewers

Darwin Grosse is the Director of Education and Services with Cycling '74, the developer of the Max media programming system. He is also an Adjunct Professor at the University of Denver, and teaches sonic art, programming, and hardware interface in the Emerging Digital Practices department.

Pradumn Joshi is currently pursuing his Bachelor's degree in Electrical Engineering from NIT Surat. He is an avid elocutionist and debate enthusiast, and is also interested in economics, freelance writing, and Western music. His area of technical expertise lies in open source hardware development and embedded systems.

Phillip Mayhew is a Bachelor of Science in Computer Science from North Carolina State University. He is the Founder and Managing Principal of Rextency Technologies LLC based in Statesville, North Carolina. His primary expertise is in software application performance testing and monitoring.

Glenn D. Reuther's own personal journey and fascination began with music technology during the 1970s with private lessons in "Electronic Music Theory and Acoustic Physics". He then attended Five Towns College of Music in NY and has been a home studio operator since 1981, playing multiple instruments and designing a few devices for his studio configuration.

Since then, he has spent several years with Grumman Aerospace as a Ground and Flight Test Instrumentation Technician, before moving through to the IT field. Beginning with an education in Computer Operations and Programming, he went on to work as network and system engineer having both Microsoft and Novell certifications. After over 10 years at the University of Virginia as Sr. Systems Engineer, he spends much of his spare time working with the current state of music technology. His website is `http://lico.drupalgardens.com`.

He is also the author of "One Complete Revelation", a photo journal of his nine-month trek throughout Europe during the early 90s.

I would like to thank the author for his friendship, and I would also like to thank my wonderful wife Alice and son Glenn for their patience, understanding, and support during the editing process of this book.

Steve Spence has been a veteran of the IT industry for more than 20 years, specializing in network design and security. Currently he designs microcontroller-based process controls and database-driven websites. He lives off grid and teaches solar and wind power generation workshops. He's a former firefighter and rescue squad member, and a current Ham Radio operator.

In the past, he's been a technical reviewer of various books on alternative fuels (From the Fryer to the Fuel Tank, Joshua Tickell) and authored DIY alternative energy guides.

www.PacktPub.com

Support files, eBooks, discount offers and more

You might want to visit www.PacktPub.com for support files and downloads related to your book.

Did you know that Packt offers eBook versions of every book published, with PDF and ePub files available? You can upgrade to the eBook version at www.PacktPub.com and as a print book customer, you are entitled to a discount on the eBook copy. Get in touch with us at service@packtpub.com for more details.

At www.PacktPub.com, you can also read a collection of free technical articles, sign up for a range of free newsletters and receive exclusive discounts and offers on Packt books and eBooks.

http://PacktLib.PacktPub.com

Do you need instant solutions to your IT questions? PacktLib is Packt's online digital book library. Here, you can access, read and search across Packt's entire library of books.

Why Subscribe?

- Fully searchable across every book published by Packt
- Copy and paste, print and bookmark content
- On demand and accessible via web browser

Free Access for Packt account holders

If you have an account with Packt at www.PacktPub.com, you can use this to access PacktLib today and view nine entirely free books. Simply use your login credentials for immediate access.

Table of Contents

Preface

Our futuristic world is full of smart and connected devices. Do-it-yourself communities have always been fascinated by the fact that each one could design and build its own smart system, dedicated or not, for specific tasks. From small controllers switching on the lights when someone is detected to a smart sofa sending e-mails when we sit on them, cheap electronics projects have become more and more easy to create and, for contributing to this, we all have to thank the team, who initiated the Arduino project around 2005 in Ivrea, Italy.

Arduino's platform is one of the most used open source hardware in the world. It provides a powerful microcontroller on a small printed circuit board with a very small form factor. Arduino users can download the Arduino **Integrated Development Environment (IDE)** and code their own program using the C/C++ language and the Arduino Core library that provides a lot of helpful functions and features.

With *C Programming for Arduino*, users will learn enough of C/C++ to be able to design their own hardware based on Arduino. This is an all-in-one book containing all the required theory illustrated with concrete examples. Readers will also learn about some of the main interaction design and real-time multimedia frameworks such as Processing and the Max 6 graphical programming framework.

C Programming for Arduino will teach you the famous "learning-by-making" way of work that I try to follow in all of my courses from Max 6 to Processing and Ableton Live.

Lastly, *C Programming for Arduino* will open new fields of knowledge by looking at the input and output concept, communication and networking, sound synthesis, and reactive systems design. Readers will learn the necessary skills to be able to continue their journey by looking at the modern world differently, not only as a user but also as a real maker.

For more details, you can visit my website for the book at
http://cprogrammingforarduino.com/.

What this book covers

Chapter 1, Let's Plug Things, is your first contact with Arduino and microcontroller programming. We will learn how to install the Arduino Integrated Development Environment on our computer and how to wire and test the development toolchain to prepare the further study.

Chapter 2, First Contact with C, covers the relation between the software and the hardware. We will introduce the C language, understand how we can compile it, and then learn how to upload our programs on the Arduino Board. We will also learn all the steps required to transform a pure idea into firmware for Arduino.

Chapter 3, C Basics – Making You Stronger, enters directly into the C language. By learning basics, we learn how to read and write C programs, discovering the datatype, basic structures, and programming blocks.

Chapter 4, Improving Programming with Functions, Math, and Timing, provides the first few keys to improve our C code, especially by using functions. We learn how to produce reusable and efficient programming structures.

Chapter 5, Sensing with Digital Inputs, introduces digital inputs to Arduino. We will learn how to use them and understand their inputs and outputs. We will also see how Arduino uses electricity and pulses to communicate with everything.

Chapter 6, Sensing the World – Feeling with Analog Inputs, describes the analog inputs of Arduino through different concrete examples and compares them to digital pins. Max 6 frameworks are introduced in this chapter as one of the ideal companions for Arduino.

Chapter 7, Talking over Serial, introduces the communication concept, especially by teaching about Serial communication. We will learn how to use the Serial communication console as a powerful debugging tool.

Chapter 8, Designing Visual Output Feedback, talks about the outputs of Arduino and how we can use them to design visual feedback systems by using LEDs and their systems. It introduces the powerful PWM concept and talks about LCD displays too.

Chapter 9, Making Things Move and Creating Sounds, shows how we can use the Arduino's outputs for movement-related projects. We talk about motors and movement and also about air vibration and sound design. We describe some basics about digital sound, MIDI, and the OSC protocol, and have fun with a very nice PCM library providing the feature of reading digitally encoded sound files from Arduino itself.

Chapter 10, Some Advanced Techniques, delivers many advanced concepts, from data storage on EEPROM units, and communication between multiple Arduino boards, to the use of GPS modules. We will also learn how to use our Arduino board with batteries, play with LCD displays, and use the VGA shield to plug the microcontroller to a typical computer screen.

Chapter 11, Networking, introduces the network concepts we need to understand in order to use our Arduino on Ethernet, wired or wireless networks. We will also use a powerful library that provides us a way to tweet messages directly by pushing a button on our Arduino, without using any computer.

Chapter 12, Playing with the Max 6 Framework, teaches some tips and techniques we can use with the Max 6 graphical programming framework. We will completely describe the use of the Serial object and how to parse and select data coming from Arduino to the computer. We will design a small sound-level meter using both real LEDs and Max 6 and finish by designing a Pitch shift sound effect controlled by our own hand and a distance sensor.

Chapter 13, Improving Your C Programming and Creating Libraries, is the most advanced chapter of the book. It describes some advanced C concepts that can be used to make our code reusable, more efficient, and optimized, through some nice and interesting real-world examples.

Appendix provides us with details of data types in C programming language, operator precedence in C and C++, important Math functions, Taylor series for calculation optimizations, an ASCII table, instructions for installing a library, and a list of components' distributors.

Appendix can be downloaded from `http://www.packtpub.com/sites/default/files/downloads/7584OS_Appendix.pdf`.

What you need for this book

If you want to take benefits of each example in this book, the following software is required:

- The Arduino environment (free, `http://arduino.cc/en/main/software`). This is required for all operations related to Arduino programming.

- Fritzing (free, `http://fritzing.org/download`). This is an open source environment that helps us design circuits.

- Processing (free, `http://processing.org/download`). This is an open source framework for rapid prototyping using Java. Some examples use it as a communication partner for our Arduino boards.

- The Max 6 framework (trial version of 30 days, `http://cycling74.com/downloads`). This framework is a huge environment that is used in this book too.

Some other libraries are also used in this book. Every time they are needed, the example description explains where to download them from and how to install them on our computer.

Who this book is for

This book is for people who want to master do-it-yourself electronic hardware making with Arduino boards. It teaches everything we need to know to program firmware using C and how to connect the Arduino to the physical world, in great depth. From interactive-design art school students to pure hobbyists, from interactive installation designers to people wanting to learn electronics by entering a huge and growing community of physical computing programmers, this book will help everyone interested in learning new ways used to design smart objects, talking objects, efficient devices, and autonomous or connected reactive gears.

This book opens new vistas of learning-by-making, which will change readers' lives.

Conventions

In this book, you will find a number of styles of text that distinguish between different kinds of information. Here are some examples of these styles, and an explanation of their meaning.

Code words in text are shown as follows: "We can include other contexts through the use of the `include` directive."

A block of code is set as follows:

```
[default]
exten => s,1,Dial(Zap/1|30)
exten => s,2,Voicemail(u100)
exten => s,102,Voicemail(b100)
exten => i,1,Voicemail(s0)
```

When we wish to draw your attention to a particular part of a code block, the relevant lines or items are set in bold:

```
[default]
exten => s,1,Dial(Zap/1|30)
exten => s,2,Voicemail(u100)
```

```
exten => s,102,Voicemail(b100)
exten => i,1,Voicemail(s0)
```

Any command-line input or output is written as follows:

```
# cp /usr/src/asterisk-addons/configs/cdr_mysql.conf.sample
   /etc/asterisk/cdr_mysql.conf
```

New terms and **important words** are shown in bold. Words that you see on the screen, in menus or dialog boxes for example, appear in the text like this: "clicking the **Next** button moves you to the next screen."

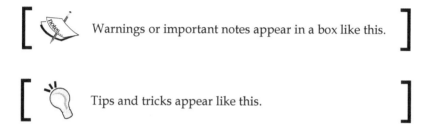

Warnings or important notes appear in a box like this.

Tips and tricks appear like this.

Reader feedback

Feedback from our readers is always welcome. Let us know what you think about this book—what you liked or may have disliked. Reader feedback is important for us to develop titles that you really get the most out of.

To send us general feedback, simply send an e-mail to feedback@packtpub.com, and mention the book title via the subject of your message.If there is a topic that you have expertise in and you are interested in either writing or contributing to a book, see our author guide on www.packtpub.com/authors.

Customer support

Now that you are the proud owner of a Packt book, we have a number of things to help you to get the most from your purchase.

Downloading the example code

You can download the example code files for all Packt books you have purchased from your account at http://www.packtpub.com. If you purchased this book elsewhere, you can visit http://www.packtpub.com/support and register to have the files e-mailed directly to you.

Errata

Although we have taken every care to ensure the accuracy of our content, mistakes do happen. If you find a mistake in one of our books—maybe a mistake in the text or the code—we would be grateful if you would report this to us. By doing so, you can save other readers from frustration and help us improve subsequent versions of this book. If you find any errata, please report them by visiting http://www.packtpub. com/submit-errata, selecting your book, clicking on the **errata submission form** link, and entering the details of your errata. Once your errata are verified, your submission will be accepted and the errata will be uploaded on our website, or added to any list of existing errata, under the Errata section of that title. Any existing errata can be viewed by selecting your title from http://www.packtpub.com/support.

Piracy

Piracy of copyright material on the Internet is an ongoing problem across all media. At Packt, we take the protection of our copyright and licenses very seriously. If you come across any illegal copies of our works, in any form, on the Internet, please provide us with the location address or website name immediately so that we can pursue a remedy.

Please contact us at copyright@packtpub.com with a link to the suspected pirated material.

We appreciate your help in protecting our authors, and our ability to bring you valuable content.

Questions

You can contact us at questions@packtpub.com if you are having a problem with any aspect of the book, and we will do our best to address it.

1
Let's Plug Things

Arduino is all about plugging things. We are going to do that in a couple of minutes after we have learned a bit more about microcontrollers in general and especially the big and amazing Arduino family. This chapter is going to teach you how to be totally ready to code, wire, and test things with your new hardware friend. Yes, this will happen soon, very soon; now let's dive in!

What is a microcontroller?

A **microcontroller** is an **integrated circuit** (**IC**) containing all main parts of a typical computer, which are as follows:

- Processor
- Memories
- Peripherals
- Inputs and outputs

The **processor** is the brain, the part where all decisions are taken and which can calculate.

Memories are often both spaces where both the core inner-self program and the user elements are running (generally called **Read Only Memory** (**ROM**) and **Random Access Memory** (**RAM**)).

I define peripherals by the self-peripherals contained in a global board; these are very different types of integrated circuits with a main purpose: to support the processor and to extend its capabilities.

Inputs and outputs are the ways of communication between the world (around the microcontroller) and the microcontroller itself.

The very first single-chip processor was built and proposed by Intel Corporation in 1971 under the name **Intel 4004**. It was a 4-bit **central processing unit** (**CPU**).

Since the 70s, things have evolved a lot and we have a lot of processors around us. Look around, you'll see your phone, your computer, and your screen. Processors or microprocessors drive almost everything.

Compared to microprocessors, microcontrollers provide a way to reduce power consumption, size, and cost. Indeed, microprocessors, even if they are faster than processors embedded in microcontrollers, require a lot of peripherals to be able to work. The high-level of integration provided by a microcontroller makes it the friend of embedded systems that are car engine controller, remote controller of your TV, desktop equipment including your nice printer, home appliances, games of children, mobile phones, and I could continue…

There are many families of microcontrollers that I cannot write about in this book, not to quote **PICs** (http://en.wikipedia.org/wiki/PIC_microcontroller) and **Parallax SX** microcontroller lines. I also want to quote a particular music hardware development open source project: **MIDIbox** (PIC-, then STM32-based, check http://www.ucapps.de). This is a very strong and robust framework, very tweakable. The Protodeck controller (http://julienbayle.net/protodeck) is based on MIDIbox.

Now that you have understood you have a whole computer in your hands, let's specifically describe Arduino boards!

Presenting the big Arduino family

Arduino is an open source (http://en.wikipedia.org/wiki/Open_source) singleboard-based microcontroller. It is a very popular platform forked from the **Wiring** platform (http://www.wiring.org.co/) and firstly designed to popularize the use of electronics in interaction design university students' projects.

My Arduino MEGA in my hand

It is based on the Atmel AVR processor (http://www.atmel.com/products/microcontrollers/avr/default.aspx) and provides many inputs and outputs in only one self-sufficient piece of hardware. The official website for the project is http://www.arduino.cc.

The project was started in Italy in 2005 by founders Massimo Banzi and David Cuartielles. Today it is one of the most beautiful examples of the open source concept, brought to the hardware world and being often used only in the software world.

We talk about Arduino family because today we can count around 15 boards 'Arduino-based', which is a funny meta-term to define different type of board designs all made using an Atmel AVR processor. The main differences between those boards are the:

- Type of processor
- Number of inputs and outputs
- Form factor

Some Arduino boards are a bit more powerful, considering calculation speed, some other have more memory, some have a lot of inputs/outputs (check the huge Arduino Mega), some are intended to be integrated in more complex projects and have a very small form factor with very few inputs and outputs… as I used to tell my students *each one can find his friend in the Arduino family*. There are also boards that include peripherals like Ethernet Connectors or even Bluetooth modules, including antennas.

The magic behind this family is the fact we can use the same **Integrated Development Environment (IDE)** on our computers with any of those boards (http://en.wikipedia.org/wiki/Integrated_development_environment). Some bits need to be correctly setup but this is the very same software and language we'll use:

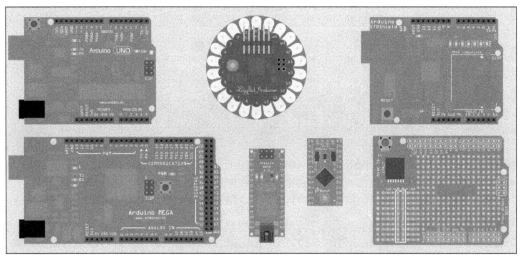

Some notable Arduino family members: Uno R3, LilyPad, Arduino Ethernet, Arduino Mega, Arduino Nano, Arduino Pro, and a prototyping shield

A very nice but non-exhaustive reference page about this can be found at http://arduino.cc/en/Main/Hardware.

I especially want you to check the following models:

- **Arduino Uno** is the basic one with a replaceable chipset
- **Arduino Mega**, 2560 provides a bunch of inputs and outputs
- **Arduino LilyPad**, is wearable as clothes
- **Arduino Nano**, is very small

Throughout this book I'll use an Arduino Mega and Arduino Uno too; but don't be afraid, when you've mastered Arduino programming, you'll be able to use any of them!

About hardware prototyping

We can program and build software quite easily today using a lot of open source frameworks for which you can find a lot of helpful communities on the Web. I'm thinking about **Processing** (Java-based, check http://processing.org), and **openFrameworks** (C++-based, check http://www.openframeworks.cc), but there are many others that sometimes use very different paradigms like graphical programming languages such as **Pure Data** (http://puredata.info), **Max 6** (http://cycling74.com/products/max/), or **vvvv** (http://vvvv.org) for Windows.

Because we, the makers, are totally involved in do-it-yourself practices, we all want and need to build and design our own tools and it often means hardware and electronics tools. We want to extend our computers with sensors, blinking lights, and even create standalone gears.

Even for testing very basic things like blinking a **light emitting diode** (LED), it involves many elements from supplying power to chipset low-level programming, from resistors value calculations to voltage-driven quartz clock setup. All those steps just gives headache to students and even motivated ones can be put off making just a first test.

Arduino appeared and changed everything in the landscape by proposing an inexpensive and all-included solution (we have to pay $30 for the Arduino Uno R3), a cross-platform toolchain running on Windows, OS X, and Linux, a very easy high-level C language and library that can also tweak the low-level bits, and a totally extensible open source framework.

Indeed, with an all-included small and cute board, an USB cable, and your computer, you can learn electronics, program embedded hardware using C language, and blink your LED.

Hardware prototyping became (almost) as easy as software prototyping because of the high level of integration between the software and the hardware provided by the whole framework.

One of the most important things to understand here is the prototyping cycle.

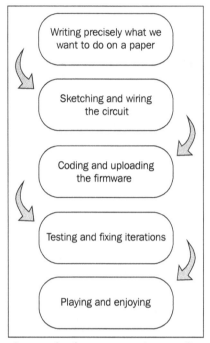

One easy hardware prototyping steps list

From our idea to our final render, we usually have to follow these steps.

If we want to make that LED blink, we have to define several blinking characteristics for instance. It will help to precisely define the project, which is a key to success.

Then we'll have to sketch a schematic with our Arduino board and our LED; it will dig the question, "How are they connected together?"

The firmware programming using C language can directly be started after we have sketched the circuit because, as we'll see later, it is directly related to the hardware. This is one of the strong powers of Arduino development. You remember? The board design has been designed only to make us think about our project and not to confuse us with very low-level abstract learning bits.

The upload step is a very important one. It can provide us a lot of information especially in case of further troubleshooting. We'll learn that this step doesn't require more than a couple of clicks once the board is correctly wired to our computer.

Then, the subcycle test and fix will occur. We'll learn by making, by testing, and it means by failing too. It is an important part of the process and it will teach you a lot. I have to confess something important here: at the time when I first began my **bonome** project (http://julienbayle.net/bonome), an RGB monome clone device, I spent two hours fixing a reverse wired LED matrix. Now, I know them very well because I failed one day.

The last step is the coolest one. I mentioned it because we have to keep in our mind the final target, the one that will make us happy in the end; it is a secret to succeed!

Understanding Arduino software architecture

In order to understand how to make our nice Arduino board work exactly as we want it to, we have to understand the global software architecture and the toolchain that we'll be using quite soon.

Take your Arduino board in hand. You'll see a rectangle-shaped IC with the word ATMEL written on the top; this is the processor.

This processor is the place that will contain the entire program that we'll write and that will make things happen.

When we buy (check *Appendix G, List of Components' Distributors*, and this link: http://arduino.cc/en/Main/Buy) an Arduino, the processor, also named *chipset*, is preburnt. It has been programmed by careful people in order to make our life easier. The program already contained in the chipset is called the **bootloader** (http://en.wikipedia.org/wiki/Booting). Basically, it takes care of the very first moment of awakening of the processor life when you supply it some power. But its major role is the load of our firmware (http://en.wikipedia.org/wiki/Firmware), I mean, our precious compiled program.

Let's have a look at a small diagram for better understanding:

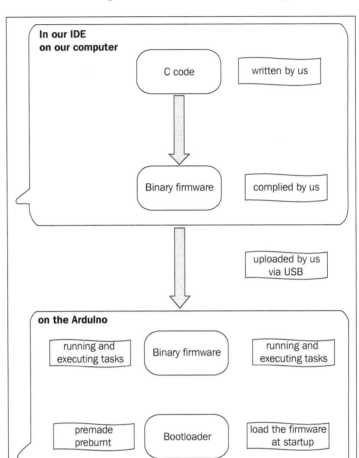

I like to define it by saying that *the bootloader is the hardware's software and the firmware is the user's software*. Indeed, it also has some significance because memory spaces in the chipset are not equal for write operations (within a specific hardware which we'll discuss in the future sections of this book). Using a **programmer**, we cannot overwrite the bootloader (which is safer at this point of our reading) but only the firmware. This will be more than enough even for advanced purposed, as you'll see all along the book.

Not all Arduino boards' bootloaders are equivalent. Indeed, they have been made to be very specific to the hardware part, which provides us more abstraction of the hardware; we can focus on higher levels of design because the bootloader provides us services such as firmware upload via USB and serial monitoring.

Let's now download some required software:

- **FTDI USB drivers**: http://www.ftdichip.com/Drivers/VCP.htm
- **Arduino IDE**: http://arduino.cc/en/Main/Software
- **Processing**: http://processing.org/download/

Processing is used in this book but isn't necessary to program and use Arduino boards.

What is the Arduino's toolchain?

Usually, we call Arduino's **toolchain** a set of software tools required to handle all steps from the C code we are typing in the Arduino IDE on our computer to the firmware uploaded on the board. Indeed, the C code you type has to be prepared before the compilation step with avr-gcc and avr-g++ compilers. Once the resulting object's files are linked by some other programs of the toolchain, into usually only one file, you are done. This can later be uploaded to the board. There are other ways to use Arduino boards and we'll introduce that in the last chapter of this book.

Installing Arduino development environment (IDE)

Let's find the compressed file downloaded from http://arduino.cc/en/Main/Software in the previous part and let's decompress it on our computer.

Whatever the platform, the IDE works equally and even if I'll describe some specific bits of three different platforms, I'll only describe the use of the IDE and show screenshots from OS X.

Installing the IDE

There isn't a typical installation of the IDE because it runs into the **Java Virtual Machine**. This means you only have to download it, to decompress it somewhere on your system, and then launch it and JAVA will execute the program. It is possible to use only the **CLI** (**command-line interface**, the famous g33ks window in which you can type the command directly to the system) to build your binaries instead of the graphical interface, but at this point, I don't recommend this.

Usually, Windows and OS X come with Java installed. If that isn't the case, please install it from the `java.com` website page at `http://www.java.com/en/download/`.

On Linux, the process really depends on the distribution you are using, so I suggest to check the page `http://www.arduino.cc/playground/Learning/Linux` and if you want to check and install all the environment and dependencies from sources, you can also check the page `http://www.arduino.cc/playground/Linux/All`.

How to launch the environment?

In Windows, let's click on the `.exe` file included in the uncompressed folder. On OS X, let's click on the global self-contained package with the pretty Arduino logo. On Linux, you'll have to start the Arduino script from the GUI or by typing in the CLI.

You have to know that using the IDE you can do everything we will make in this book.

What does the IDE look like?

The IDE provides a graphical interface in which you can write your code, debug it, compile it, and upload it, basically.

The famous Blink code example opened in the Arduino IDE

There are six icons from left to right that we have to know very well because we'll use them every time:

- **Verify** (check symbol): This provides code checking for errors
- **Upload** (right-side arrow): This compiles and uploads our code to the Arduino board
- **New** (small blank page): This creates a new blank sketch
- **Open** (up arrow): This opens a list of all sketches already existing in our sketchbook
- **Save** (down arrow): This saves our sketch in our sketchbook
- **Serial Monitor** (small magnifying glass): This provides the serial monitoring

Each menu item in the top bar provides more options we will discover progressively all throughout this book.

However, the **Tools** menu deserves closer attention:

- **Auto Format**: This provides code formatting with correct and standard indentations
- **Archive Sketch**: This compresses the whole current sketch with all files
- **Board**: This provides a list of all boards supported
- **Serial Port**: This provides a list of all serial devices on the system
- **Programmer**: This provides a list of all programmer devices supported and used in case of total reprogramming of the AVR chipset

- **Burn Bootloader**: This is the option used when you want to overwrite (or even write) a new bootloader on your board.

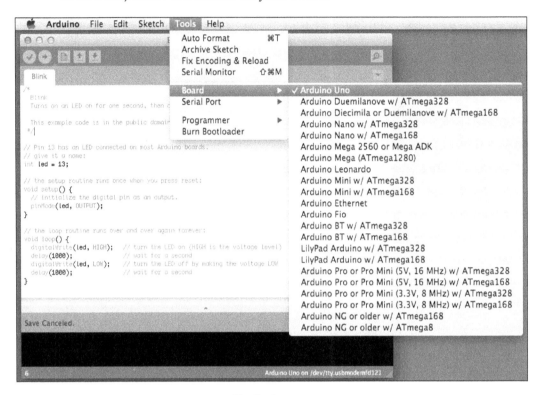

The Tools menu

The preferences dialog is also a part we have to learn about right now. As usual, the preferences dialog is a place where we don't really need to go often but only for changing global parameters of the IDE. You can choose the sketchbook location and the Editor language in this dialog. You can also change a couple of bits like automatic check-up of IDE updates at start up or Editor font size.

The sketchbook concept will make our life easier. Indeed, the sketchbook is a folder where, basically, all your sketches will go. On my personal point of view, it is very precious to use it like this because it really organizes things for you and you can retrieve your pieces of code easier. Follow me there; you'll thank me later.

When we start a sketch from scratch, we basically type the code, verify it, upload it, and save it. By saving it, the first time, the IDE creates a folder in which it will put all the files related to our current sketch. By clicking on the sketch file inside this folder, the Arduino IDE will open and the related code will be displayed in the edit/typing part of the window.

We are almost done!

Let's install the drivers of the Arduino USB interface on our system.

Installing Arduino drivers

Arduino boards provide an USB interface. Before we plug the USB cable and link the board to our computer, we have to install specific drivers in the latter.

There is a huge difference between Windows and OS X here; basically, OS X doesn't require any specific drivers for Arduino Uno or even Mega 2560. If you are using older boards, you'd have to download the latest version of drivers on the FTDI website, double-click the package, then follow instructions, and finally, restart your computer.

Let's describe how it works on Windows-based systems, I mean, Windows 7, Vista, and XP.

Installing drivers for Arduino Uno R3

It is important to follow the steps mentioned next to be able to use the Arduino Uno R3 and some other boards. Please check the Arduino website for up-to-date references.

1. Plug your board in and wait for Windows to begin the driver installation process. After a few moments, the process fails.

2. Click on the Start menu, and open **Control Panel**.

3. In **Control Panel**, navigate to **System and Security**. Next, click on **System**. Once the **System** window is up, open **Device Manager**.

4. Look under **Ports (COM & LPT)**. Check the open port named **Arduino UNO (COMxx)**.

5. Right-click on **the Arduino UNO (COMxx)** port and choose the **Update Driver Software** option.

6. Next, choose the **Browse my computer for driver software** option.

7. Finally, navigate and select the Uno's driver file, named `ArduinoUNO.inf`, located in the `Drivers` folder of the Arduino software download (be careful: not the `FTDI USB Drivers` subdirectory).

8. Windows will finish the driver installation from there and everything will be fine.

Installing drivers for Arduino Duemilanove, Nano, or Diecimilla

When you connect the board, Windows should initiate the driver installation process (if you haven't used the computer with an Arduino board before).

On Windows Vista, the driver should be automatically downloaded and installed. (Really, it works!)

On Windows XP, the **Add New Hardware** wizard will open:

1. When asked **Can Windows connect to Windows Update to search for software?** select **No, not this time**. Click on **Next**.

2. Select **Install from a list or specified location (Advanced)** and click on **Next**.

3. Make sure that **Search for the best driver in these locations** is checked, uncheck **Search removable media**, check **Include this location in the search**, and browse to the **drivers/FTDI USB Drivers** directory of the Arduino distribution. (The latest version of the drivers can be found on the *FTDI* website.) Click on **Next**.

4. The wizard will search for the driver and then tell you that a **USB Serial Converter** was found. Click on **Finish**.

5. The new hardware wizard will appear again. Go through the same steps and select the same options and location to search. This time, a **USB Serial Port** will be found.

You can check that the drivers have been installed by opening **Windows Device Manager** (in the **Hardware** tab of the **System** control panel). Look for a **USB Serial Port** in the **Ports** section; that's the Arduino board.

Now, our computer can recognize our Arduino board. Let's move to the physical world a bit to join together the tangible and intangible worlds.

What is electricity?

Arduino is all about electronic, and electronic refers to electricity. This may be your first dive into this amazing universe, made of wires and voltages, including blinking LEDs and signals. I'm defining several very useful notions in this part; you can consider turning down the corner of this page and to come back as often as you need.

Here, I'm using the usual analogy of water. Basically, wires are pipes and water is electricity itself.

Voltage

Voltage is a potential difference. Basically, this difference is created and maintained by a generator. This value is expressed in Volt units (the symbol is V).

The direct analogy with hydraulic systems compare the voltage to the difference of pressure of water in two points of a pipe. The higher the pressure, the faster the water moves, for a constant diameter of pipe of course.

We'll deal with low voltage all throughout this book, which means nothing more than 5 V. Very quickly, we'll use 12 V to supply motors and I'll precise that each time we do.

When you switch on the generator of closed circuits, it produces and keeps this potential difference. Voltage is a difference and has to be measured between two points on a circuit. We use voltmeters to measure the voltage.

Current and power

Current can be compared to the hydraulic volume flow rate, which is the volumetric quantity of flowing water over a time interval.

The current value is expressed in Ampères (the symbol is A). The higher the current, the higher will be the quantity of electricity moving.

A flow rate doesn't require two points to be measured as a difference of pressure; we only need one point of the circuit to make our measurement with an equipment named Ampere meter.

In all of our applications, we'll deal with **direct current** (**DC**), which is different from **alternative current** (**AC**).

Power is a specific notion, which is expressed in Watt (the symbol is W).

Following is a mathematical relationship between voltage, current, and power:

$P = V \times I$

where, P is the power in Watt, V the voltage in V, and I the current in Ampères.

Are you already feeling better? This analogy has to be understood as a proper analogy, but it really helps to understand what we'll make a bit later.

And what are resistors, capacitors, and so on?

Following the same analogy, **resistors** are small components that slow down the flow of current. They are more resistive than any piece of wire you can use; they generally dissipate it as heat. They are two passive terminal components and aren't polarized, which means you can wire them in both directions.

Resistors are defined by their *electrical resistance* expressed in Ohms (the symbol is Ω).

There is a direct mathematical relation between voltage measured at the resistor sides, current, and resistance known as the Ohm's law:

R = V / I

where R the electrical resistance in Ohms, V the voltage in Volts, and I the current in Ampères.

For a constant value of voltage, if the resistance is high, the current is low and vice-versa. It is important to have that in mind.

On each resistor, there is a color code showing the resistance value.

There are many types of resistors. Some have a constant resistance, some others can provide different resistance values depending on physical parameters such as temperature, or light intensity for instance.

A **potentiometer** is a variable resistor. You move a slider or rotate a knob and the resistance changes. I guess you begin to understand my point…

A **capacitor** (or **condenser**) is another type of component used very often. The direct analogy is the rubber membrane put in the pipe: no water can pass through it, but water can move by stretching it.

They are also passive two-terminal components but can be polarized. Usually, small capacitors aren't.

We usually are saying that capacitors store potential energy by charging. Indeed, the rubber membrane itself stores energy while you stretch it; try to release the stretched membrane, it will find its first position.

Capacitance is the value defining each capacitor. It is expressed in Farads (the symbol is F).

We'll stop here about capacitance calculations because it involves advanced mathematics which isn't the purpose of this book. By the way, keep in mind the higher the capacitance, more will be the potential the capacitor can store.

A **diode** is again a two-terminal passive component but is polarized. It lets the current pass through it only in one direction and stop it in the other. We'll see that even in the case of direct current, it can help and make our circuits safer in some cases.

LEDs are a specific type of diode. While the current passes through them in the correct direction, they glow. This is a nice property we'll use to check if our circuit is correctly closed in a few minutes.

Transistor is the last item I'm describing here because it is a bit more complex, but we cannot talk about electronics without quoting it.

Transistors are semiconductor devices that can amplify and switch electronics signals and power, depending on how they are used. They are three-terminal components. This is the key active component of almost all modern electronics around us. Microprocessors are made of transistors and they can even contain more than 1 billion of them.

Transistors in the Arduino world are often used to drive high current, which couldn't pass through the Arduino board itself without burning it. In that case, we basically use them as analogue switches. When we need them to close a circuit of high currents to drive a motor for instance, we just drive one of their three terminals with a 5 V coming from the Arduino and the high current flows through it as if it had closed a circuit. In that case, it extends the possibilities of the Arduino board, making us able to drive higher currents with our little piece of hardware.

Wiring things and Fritzing

With the previous analogy, we can understand well that a circuit needs to be closed in order to let the current flow.

Circuits are made with wires, which are basically conductors. A conductor is a matter with a resistance near to zero; it lets the current flow easily. Metals are usually good conductors. We often use copper wires.

In order to keep our wiring operations easy, we often use pins and headers. This is a nice way to connect things without using a soldering iron each time!

By the way, there are many ways to wire different components together. For our prototyping purpose, we won't design printed circuit board or even use our soldering iron; we'll use breadboards!

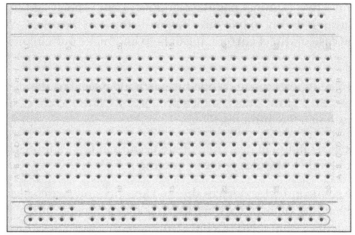

A breadboard with its buses blue and red and its numerous perforations

Breadboards are the way to rapid prototyping and this is the way to go here.

Basically, breadboards consists of a piece of plastic with many perforations in which there are small pieces of conductors allowing to connect wires and components' leads inside.

The distance between two perforations is 2.54 mm (equal to 0.1") that is a standard; for instance, dual in-line package integrated circuits' leads are all separated by this particular distance and thus, you can even put IC on breadboards.

As we saw on the previous screenshot, there are buses and terminals strips.

Buses are series of five perforations in the central part and put in column for which the underlying conductors are connected. I have surrounded one bus with a green stroke.

Terminals are special buses usually used for power supplying the circuit and appear in between blue and red lines. Usually, we use blue for ground lines and red for voltage source (5 V or 3.3 V in some cases). A whole line of terminals has its perforations all connected, providing voltage source and ground easily available on all the breadboard without having to use a lot of connection to the Arduino. I surrounded 2 of the 4 terminals with red and blue strokes.

Breadboards provide one of the easiest ways of prototyping without soldering. It also means you can use and reuse your breadboards throughout the years!

What is Fritzing?

I discovered the open source **Fritzing** project (http://fritzing.org) when I needed a tool to make my first master classes slideshows schematic around the Protodeck controller (http://julienbayle.net/protodeck) I built in 2010.

Fritzing is defined as *an open source initiative to support designers, artists, researchers and hobbyists to work creatively with interactive electronics.* It sounds as if it had been made for us, doesn't it?

You can find the Fritzing's latest versions at http://fritzing.org/download/.

Basically, with Fritzing, you can design and sketch electronic circuits. Because there are many representations of electronic circuits, this precious tool provides two of the classic ones and a PCB design tool too.

Considering the first practical work we are going to do, you have to take your breadboard, your Arduino, and wire the lead and the resistor exactly as it is shown in the next screenshot:

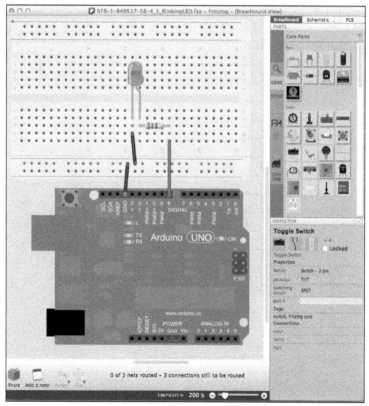

The breadboard view showing our first circuit

The *breadboard view* is the one that looks the most like what we have in front of us on the table. You represent all wires and you connect a virtual breadboard to your Arduino and directly plug components.

The magic lies in the fact that the schematic is automatically build while you are sketching in the breadboard view. And it works both ways! You can make a schematic, and Fritzing connect components in the breadboard view. Of course, you'd probably have to place the part in a more convenient or aesthetical way, but it works perfectly fine. Especially, the **Autorouter** helps you with making all wires more linear and simple.

In the next screenshot, you can see the same circuit as before, but shown in the *schematic view*:

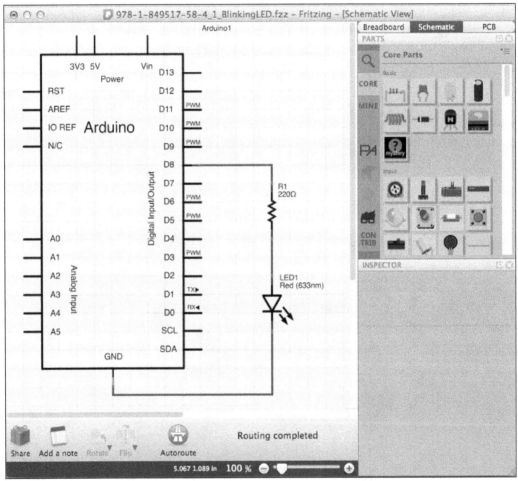

The schematic view representing the circuit diagram

There are a lot of components already designed especially for Fritzing and you can even create yours quite easily. The page to visit for this purpose is http://fritzing.org/parts/.

The native library contains all parts required in all schematics of this book from all Arduino boards, to any discrete components and IC too. Indeed, all schematics of this book have been made using Fritzing!

Now that you know how to wire things without any soldering iron, and how to quietly sketch and check things on your computer before you do it for real on your desktop, let's learn a bit about power supply.

Power supply fundamentals

We learned a bit more about electricity before, but how can I supply all my circuits in real life?

Arduino boards can be supplied in three different ways:

- By our computer via the USB cable (5 V is provided)
- By a battery or a direct external **Power Supply Unit** (**PSU**) / Adapter
- By attaching a regulated 5 V to the +5 V pin

The USB cable contains four cables: two for data communication purpose and two for power supply. Those latter are basically used to supply Arduino when you are connecting it to the computer via USB.

USB is a special communication bus that provides 5 V but no more than 500 mA. (0.5 A) It means we have to use another supply source in special projects where we need a lot of LED, motors, and other devices that drive a lot of current.

What adapter can I use with my Arduino?

Arduino Uno and Mega can be directly supplied by DC Adapter but this one *has* to respect some characteristics:

- The output voltage should be between 9 V and 12 V
- It should be able to drive at least 250 mA of current
- It must have a 2.1 mm power plug with center positive

Usually, if you ask yourself about the fact whether to use an adapter or not, it means you need more current than the USB's 500 mA (Practically, ask yourself this question whether you need around 400 mA).

Using USB or the 2.1 mm power plug with an adapter are the safest ways to use Arduino boards for many reasons. The main one is the fact that those two sources are (hopefully) clean, which means they deliver a regulated voltage.

However, you have to change something on the board if you want to use one or the other source: a jumper has to be moved to the right position:

On the left, the jumper is set to USB power supply and on the right, it is set to external power supply

Usually, an idle Arduino board drains around 100 mA and, except in specified cases (see *Chapter 9, Making Things Move and Creating Sounds*), we'll use the USB way of supply. This is what you have to do now: plug in the USB cable both in the Arduino and your computer.

Launch the Arduino IDE too, and let's move further to the hardware *Hello World* of our system, I call that the *Hello LED*!

Hello LED!

If your Arduino doesn't contain any firmware, the LED probably does nothing. If you check the built-in LED on the Arduino board itself, that one should blink.

Let's take the control over our external cute LED plugged in the breadboard right now.

What do we want to do exactly?

If you remember correctly, this is the first question we have to ask. Of course, we bypassed this step a bit especially about the hardware part because I had to explain things while you were wiring, but let's continue the prototyping process explained in part by checking the code and uploading it.

We want to make our LED blink. But what blink speed ? How much time? Let's say we want to make it blink every 250 ms with a one second pause between the blinks. And we want to do that infinitely.

If you check the schematic, you can understand that the LED is put between the ground, and the line to the digital output pin number 8.

There is a resistor and you now know that it can consume a bit of energy by resisting to the current flowing to the LED. We can say the resistor protects our LED.

In order to make the LED light up, we have to create a flow of current. Sending +5 V to the digital output number 8 can do this. That way, there will be a potential difference at the two leads of the LED, driving it to be lighted. But the digital output shouldn't be at +5 V at each time. We have to control the moment when it will provide this voltage. Still okay?

Let's summarize what we have to do:

1. Put the 5 V to the digital output 8 during 250ms.
2. Stop to drive the digital output 8 during 1s.
3. Restart this every time the Arduino is powered

How can I do that using C code?

If you followed the previous page correctly, you already have your Arduino board wired to the computer via your USB cable on one side, and wired to the breadboard on the other side.

Now, launch your Arduino IDE.

Start with a new blank page

If you already tested your IDE by loading some examples, or if you already wrote some piece of code, you have to click on the *New icon* in order to load a blank page, ready to host our `Blink250ms` code:

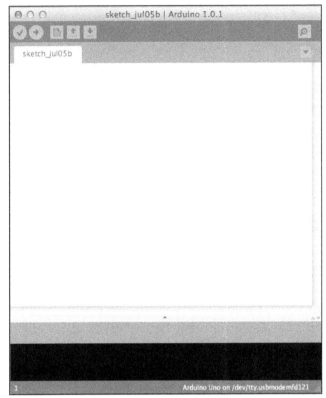

A nice and attractive blank page

Setting up the environment according the board we are using

The IDE has to know with which board it will have to communicate. We will do it in the following steps:

1. Go to the **Tools** menu and choose the correct board. The first one is **Arduino Uno**:

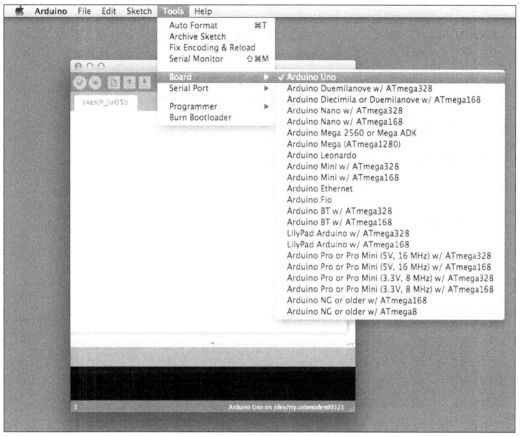

Choose the board you are using

2. Once we have done that, we have to choose the correct serial port. Go to the **Tools** menu again and choose the correct serial port:

   ° On OS X, the correct one begins with **/dev/tty.usbmodem** for both Uno and Mega 2560 and with **/dev/tty.usbserial** for older boards.

   ° On Windows, the correct port is usually **COM3** (**COM1** and **COM2** are often reserved by the operating system). By the way, it can also be **COM4**, **COM5**, or whatever else. To be sure, please check the device manager.

° On Linux, the port is usually **/dev/ttyUSB0**:

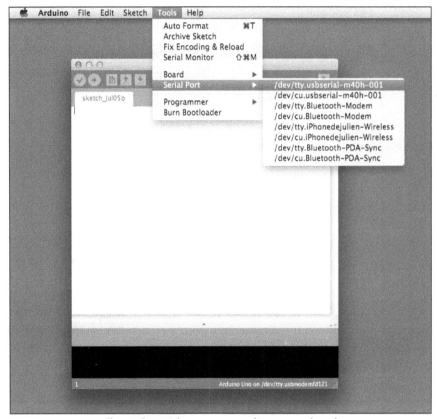

Choose the serial port corresponding to your board

Now, our IDE can talk to our board. Let's push the code now.

Let's write the code

The following is the complete code. You can find it in the zip file in the `Chapter01/` `Blink250ms/` folder:

Downloading the example code

You can download the example code files for all Packt books you have purchased from your account at `http://www.packtpub.com`. If you purchased this book elsewhere, you can visit `http://www.packtpub. com/support` and register to have the files e-mailed directly to you.

```
/*
  Blink250ms Program
  Turns a LED connected to digital pin 8 on for 250ms, then off for
1s, infinitely.
```

```
   Written by Julien Bayle, this example code is Creative Commons CC-
   BY-SA
   */

// Pin 8 is the one connected to our LED
int ledPin = 8;                  // ledPin is an integer variable
initialized at 8

// --------- the setup routine runs once when you power up the board
or push the reset switch
void setup() {
   pinMode(ledPin, OUTPUT);    // initialize the digital pin as an
output because we want it to source a current
}

// --------- the loop routine runs forever
void loop() {
   digitalWrite(ledPin, HIGH);   // turn the LED on (HIGH is a constant
meaning a 5V voltage)
   delay(250);                   // wait for 250ms in the current state
   digitalWrite(ledPin, LOW);    // turn the LED off (LOW is a constant
meaning a 5V voltage)
   delay(1000);                  // wait for 1s in the current state
}
```

Let's comment it a bit. Indeed, we'll learn how to code our own C code in the next chapter, then I'll only describe this one and give you some small tips.

First, everything between /* and */, and everything after // are just comments. The first form is used for comments more than one line at a time, and the other one is for one line commenting only. You can write any comments like that and they won't be considered by the compiler at all. I strongly advice you to comment your code; this is another key to succeed.

Then, the first part of the code contains one variable declaration and initialization:

```
int ledPin = 8;
```

Then, we can see two particular structures between curly braces:

```
void setup() {
   pinMode(ledPin, OUTPUT);
}
void loop() {
   digitalWrite(ledPin, HIGH);
   delay(250);
   digitalWrite(ledPin, LOW);
   delay(1000);
}
```

The first one (`setup()`) is a function that is executed only one time when the Arduino board is started (or reseted); this is the place where we are telling the board that the pin where the LED is connected is an output, that is, this pin will have to drive current while activated.

The second one (`loop()`) is a function executed infinitely when the Arduino board is supplied. This is the main part of our code in which we can find the steps we wanted to light up the LED for 250 ms and switch off the LED for 1 s, repeatedly.

Let's upload the code, at last!

If you correctly followed and manipulated the hardware and the IDE as explained before, we are now ready to upload the code on the board.

Just click on the Upload button in the IDE. You'll see the TX and RX LEDs blinking a bit and … your LED on your breadboard should blink as expected. This is our very first *HELLO LED!* example and I hope you liked it.

If you want to tweak the code a bit, you can replace the following line:

```
delay(1000);
```

With the following line, for instance:

```
delay(100);
```

Now upload this new code again and see what happens.

Summary

In this chapter itself, we learnt a bit about Arduino and microcontrollers, and about electricity too. That will help us in the next chapters in which we will talk a lot about circuits.

We also installed the IDE that we will use every time while programming Arduino boards and we even tested the first piece of code. We are now able to continue our travel by learning more about the C language itself.

2
First Contact with C

In my life as a programmer, I encountered a lot of compiler-based as well as scripting languages. One of the lowest common denominators has always been the C language.

In our case, this is **embedded system programming**, which is another name for **hardware programming**; this first statement is also true.

Let's check what C programming really is and let's enter into a new world, that is, the realm of Arduino programming. We'll also use a very necessary feature called **serial monitoring**. This will help us a lot in our C learning, and you'll understand that this feature is also used in real-life projects.

An introduction to programming

The first question is, **what is a program?**

A **program** is text that you write using a programming language that contains behaviors that you need a processor to acquire. It basically creates a way of handling inputs and producing outputs according to these behaviors.

According to Wikipedia (`http://en.wikipedia.org/wiki/Computer_programming`):

> *Programming is the process of designing, writing, testing, debugging and maintaining the source code of computer programs.*

Of course, this definition is very simple and it also applies to microcontrollers, as we already know that the latter are basically a type of computers.

Designing a program is the fact you have to think about first, before you begin coding it. It generally involves writing, drawing, and making schematics of all the actions you want your processor to make for you. Sometimes, it also implies to write what we call **pseudocode**. I hope you remember that this is what we created in the previous chapter when we wanted to define precisely all the steps of our desired LED behavior.

I don't agree with a lot of people calling it *pseudocode* because it is actually more of a *real code*.

What we call *pseudocode* is something that helps a lot because it is human-readable, made of clear sentences, and is used to think and illustrate better our purpose, which is the key to success.

An example of my firmware *pseudocode*'s definition could be as follows:

1. Measure the current thermic sensor value.
2. Check if the temperature is greater than 30° C and make a sound if it is.
3. If not, light the blue LED.
4. And make those previous steps permanent in a loop.

Writing a program is typically what converts the pseudocode into real and well-formed code. It involves having knowledge of programming languages because it is the step when you really write the program. This is what we'll learn in a moment.

Testing is the obvious step when you run the program after you made some modifications to the code. It is an exciting moment when you also are a bit afraid of bugs, those annoying things that make running your program absolutely different from what you expected at first.

Debugging is a very important step when you are trying to find out why that program doesn't work well as expected. You are tracking typo errors, logic discrepancies, and global program architecture problems. You'll need to monitor things and often modify your program a bit in order to precisely trace how it works.

Maintaining the source code is the part of the program's life that helps to avoid obsolescence.

The program is working and you improve it progressively; you make it up-to-date considering hardware evolutions, and sometimes, you debug it because the user has this still undiscovered bug. This step increases the life duration of your program.

Different programming paradigms

A **paradigm** is a manner of describing something. It can either be a representation or a theoretical model of something.

Applied to programming, a programming paradigm is *a fundamental style of computer programming.*

The following are four main programming paradigms:

- Object-oriented
- Imperative
- Functional
- Logic programming

Some languages follow not one but multiple paradigms.

It is not the purpose of this book to have a debate around those, but I would add one, which can be a combination of these and which also describes a particular concept: **visual programming**. We'll discover one of the most powerful frameworks in *Chapter 6, Sensing the World – Feeling with Analog Inputs*, namely the **Max 6** framework (formerly named **Max/MSP**).

Programming style

There is no scientific or universal way to define what is the absolute best style of programming. However, I can quote six items that can help to understand what we'll try to do together all along this book in order to make good programs. We'll aim for the following:

- **Reliability**: This enables a code to handle its own generated errors while running
- **Solidity**: This provides a frame to anticipate problems on the user side (wrong inputs)
- **Ergonomics**: This helps to intuitively be able to use it with ease
- **Portability**: This is the designing of a program for a wide range of platforms
- **Maintainability**: This is the ease of modifying it even if you didn't code it yourself
- **Efficiency**: This indicates that a program runs very smoothly without consuming a lot of resources

Of course, we'll come back to them in the examples of this book, and I'm sure you'll improve your style progressively.

C and C++?

Dennis Ritchie (http://en.wikipedia.org/wiki/Dennis_Ritchie) at Bell Labs developed the C programming language from 1969 to 1973. It is often defined as a general-purpose programming language and is indeed one of the most used languages of all times. It had been used initially to design the Unix operating system (http://en.wikipedia.org/wiki/Unix) that had numerous requirements, especially high performance.

It has influenced a lot of very well known and used languages such as C++, Objective-C, Java, JavaScript, Perl, PHP, and many others.

C is to both **imperative** and **structured**. It is very appropriate for both 8-bit and 64-bit processors, for systems having not only several bytes of memory but also terabytes too, and also for huge projects involving huge teams, to the smallest of projects with a single developer.

Yes, we are going to learn a language that will open your mind to global and universal programming concepts!

C is used everywhere

Indeed, the C language provides a lot of advantages. They are as follows:

- *It is small and easy to learn.*
- *It is processor-independent* because *compilers* exist for almost all processors in the world. This independence provides something very useful to programmers: they can focus on algorithms and the application levels of their job instead of thinking about the hardware level at each row of code.
- *It is a very "low-level" high-level language.*

    This is its main strength. Dennis M. Ritchie, in his book *The C Programming Language* written with Brian W. Kernighan commented on C as:

    C is a relatively "low level" language. This characterization is not pejorative; it simply means that C deals with the same sort of objects that most computers do. These may be combined and moved about with the arithmetic and logical operators implemented by real machines.

Today, this is the only language that allows interacting with the underlying hardware engine so easily and this is the reason why the Arduino toolchain is based on C.

Arduino is programmed with C and C++

C++ can be considered as a superset of C. It means C++ brings new concepts and elements to C. Basically, C++ can be defined as C with object-oriented implementation (http://en.wikipedia.org/wiki/Object-oriented_programming), which is a higher-level feature. This is a very nice feature that brings and provides new ways of design.

We'll enter together into this concept a bit later in this book but basically, in object-oriented programs, you define structures called **classes** that are a kind of a model, and you create objects called **instances** of those classes, which have their own life at runtime and which respect and inherit the structure of the class from which they came.

Object-oriented programming (OOP) provides four properties that are very useful and interesting:

- Inheritance (classes can inherit attributes and behaviors from their parent classes)
- Data encapsulation (each instance retains its data and functions)
- Object identity (each instance is an individual)
- Polymorphism (each behavior can depend on the context)

In OOP, we define classes first and then we use specific functions called **constructors** to create instances of those classes. Imagine that a class is a map of a type of house, and the instances are all the houses built according to the map.

Almost all Arduino libraries are made using C++ in order to be easily reusable, which is one of the most important qualities in programming.

The Arduino native library and other libraries

A **programming library** is a collection of resources that are available for use by programs.

They can include different types of things, such as the following:

- Configuration data
- Help and documentation resources
- Subroutines and reusable part of code
- Classes
- Type definitions

I like to say that libraries provide a **behavior encapsulation**; you don't have to know how the behavior is made for using it but you just use it.

Libraries can be very specific, or can have a global purpose.

For instance, if you intend to design firmware that connects the Arduino to the Internet in order to grab some information from a mail server, and react by making an LED matrix blink in one way or another according to the content of the mail server's response, you have the following two solutions:

- Code the whole firmware from scratch
- Use libraries

Even if we like to code things, we are happier if we can focus on the global purpose of our designs, aren't we?

In that case, we'll try to find libraries already designed specifically for the behaviors we need. For instance, there is probably a library specifically designed for LED matrix control, and another one with a server-connection purpose.

Discovering the Arduino native library

The native library is designed for a very elementary and global purpose. It means it may not be enough, but it also means you'll use it every time in all your firmware design.

You can find it at `http://arduino.cc/en/Reference/HomePage`. This page will be familiar to you by now!

It is divided in the following three parts:

- **Structure** (from global conditional control structures to more specific ones)
- **Variables** (related to types and conversions between types)
- **Functions** (from I/O functions to math calculation ones and more)

The following steps can be used to find help directly in IDE:

1. Open your IDE.

2. Go to **File | Examples**; you'll see the following screenshot:

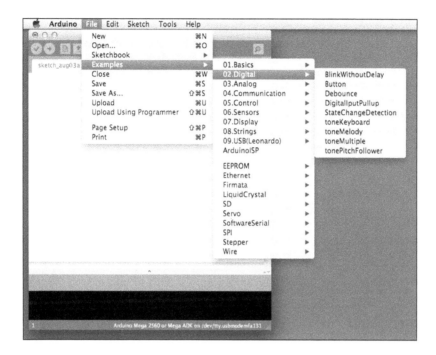

In the first part of the menu (in the preceding screenshot), you have lots of examples related to the native library only.

3. Select the **02.Digital** button.

4. A new window is displayed. Right-click on a colored keyword in the code as shown in the next screenshot:

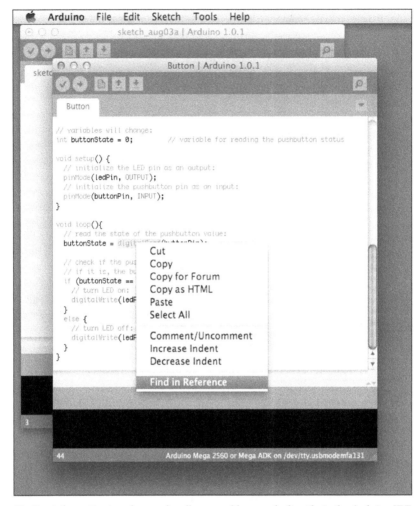

Finding information in reference for all reserved keywords directly in the Arduino IDE

5. You can see at the bottom of this contextual menu **Find in Reference**. This is a really useful tool that you are going to understand right now; click on it!

Your IDE directly called your default browser with an HTML page corresponding to the help page of the keyword on which you clicked. You can stay focused inside your IDE and go to help.

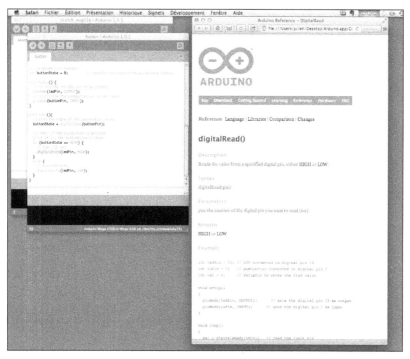

The useful local help files that are available

Other libraries included and not directly provided

The Arduino library has progressively included both necessary and useful other libraries. We have seen in the earlier chapter that the used libraries are now integrated into the *core* of the Arduino distribution, which is a bit abusive, but summarizes well the fact that they are available when you install only the Arduino IDE package.

Some very useful included libraries

- **EEPROM** provides functions and classes to read/write in hardware storage components. It is very useful to store something beyond the power state of the Arduino, that is, even when the power is off.

- **Ethernet** helps to make layer 2 and layer 3 communications over an Ethernet network.

- **Firmata** is used for serial communication.

- **SD** provides an easy way to read/write SD Cards; it is a more user-friendly alternative to the EEPROM solution.

- **Servo** helps to control servo motors.

There are a couple more libraries in the core distribution. Sometimes, new ones are included.

Some external libraries

I suggest that you check other libraries quoted and referenced on the same page at the link `http://arduino.cc/en/Reference/Libraries`.

I especially used a lot of the following libraries:

- **TLC5940**: Used to control a 16-channel, 12-bit LED controller smoothly
- **MsTimer2**: Used to trigger an action that has to be very fast and even each 1 ms (this library is also a nice hack of one of the hardware timers included in the chipset)
- **Tone**: Used to generate audible square waves

You can use *Google* to find more libraries. You will find a lot of them, but not all are equally documented and maintained. We'll see in the last chapter of this book how to create our own library, and of course how to document it nicely for both other users and ourselves.

Checking all basic development steps

We are not here together to understand the entire details of code compilation. But I want to give you a global explanation that will help you to understand better how it works under the hood. It will also help you to understand how to debug your source code and why something wouldn't work in any random case.

Let's begin by a flowchart showing the entire process.

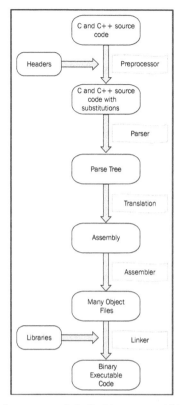

From the source code to the binary executable code

The following steps are executed to take the code from the source to the executable production stage:

1. The **C and C++ source code** is just the type of code you already wrote for the `Blink250ms` project in *Chapter 1, Let's Plug Things*.

2. **Headers** are usually included at the beginning of your code, and they refer to other files with the extension `.h` in which there are some definitions and class declarations. This kind of design, in which you have separate files for the source code (the program you are currently writing) and the headers (already made elements), provides a nice way to re-use your already written code.

3. The **Preprocessor** is a routine that basically substitutes text elements in your code, considering the *headers* and *other constants'* definitions.

4. The **Parser** prepares a file that will be translated, and that file will be assembled to produce multiple *object* files.

5. An **object** file contains machine code that is not directly executable by any hardware processor.

6. The last important step is the **linkage** made by the **linker** program. The linker takes all objects produced by the previous compilation steps and combines them into a single executable file called **program**.

7. From the source code to the object file, all processes are summarized under the name `compilation`.

8. Usually, libraries provide object files, ready to be linked by the linker. Sometimes, especially in the open source world, libraries come with source code too. This makes any changes in the library easier. In that case, the library itself would have to be compiled to produce the required object files that would be used in your global code compilation.

9. Hence, we'll define *compilation* as the whole process from the source code to the program.

I should even use and introduce another term: **cross-compilation**. Indeed, we are compiling the source code on our computer, but the final targeted processor of our resulting program (firmware) is the Arduino's processor.

Generally, we define cross-compilation as the process of compiling source code using a processor in order to make a program for another processor.

Now, let's move further and learn how we are going to test our initial pieces of C code precisely using the IDE console.

Using the serial monitor

The Arduino board itself can communicate easily using basic protocols for serial communication.

Basically, **serial communication** is the process of sending data elements over a channel, often named a **bus**. Usually, data elements are bytes, but it all depends on the implementation of the serial communication.

In serial communication, data is sent *sequentially*, one after the previous one. This is the opposite of **parallel communication**, where data are sent over more than one channel, all at the same time.

Baud rate

Because the two entities that want to communicate using serial communications have to be okay about the answer to the question "Hey, what is a word?", we have to use the same speed of transmission on both sides. Indeed, if I send `001010101010`, is it a whole word or are there many words? We have to define, for instance, that a word is four-digits long. Then, we can understand that the previous example contains three words: `0010`, `1010`, and `1010`. This involves a clock.

That clock definition is made by initializing serial communication at a particular *speed* in **baud**, also called **baud rate**.

1 baud means 1 symbol transmitted per second. A symbol can be more than one bit.

This is why we don't have to create confusion between bps, bit per second, and baud!

Serial communication with Arduino

Each Arduino board has at least one serial port. It can be used by using digital pins 0 and 1, or directly using the USB connection when you want to use serial communication with your computer.

You can check `http://arduino.cc/en/Reference/serial`.

On the Arduino board, you can read RX and TX on both digital pins 0 and 1 respectively. **TX** means transmit and **RX means** receive; indeed, the most basic serial communication requires two wires.

There are many other kinds of serial communication buses we'll describe a bit later in *Chapter 10, Some Advanced Techniques,* in the *Using I2C and SPI for LCD, LED, and other funny games* section.

 If you use serial communication on your Arduino board, you cannot use the digital pins 0 and 1.

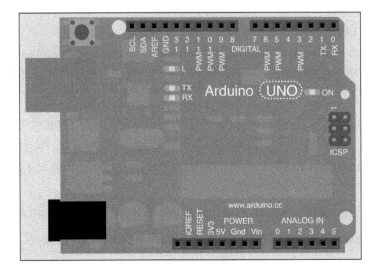

Check TX and RX on digital pins 1 and 0

Arduino IDE provides a nice serial monitor that displays all symbols sent by the board to the computer via the USB interface. It provides a lot of baud rates from 300 baud to 115,200 baud. We are going to check how to use it in the following sections.

Serial monitoring

Serial monitoring is the way of creating very basic and easy communication with our board! It means we can program it to speak to us, via the serial monitor.

If you have to debug something and the board's behavior differs from what you are expecting from it, and you want to "verify whether the problem stems from the firmware or not, you can create some routines that will write messages to you. These messages are called **traces**. Traces can be totally necessary for debugging source code.

Traces will be described in detail in the next chapter.

Making the Arduino talk to us

Imagine that you have followed carefully the `Blink250ms` project, everything is wired correctly, you double-checked that, and the code seems okay too, but it doesn't work.

Our LED isn't blinking at all. How to be sure that the `loop()` structure of your code is correctly running? We'll modify the code a bit in order to trace its steps.

Adding serial communication to Blink250ms

Here, in the following code, we'll add serial communication for the LED to blink every 250 ms:

1. Open your previous code.

2. Use **Save As** to create another project under the name `TalkingAndBlink250ms`.

 It is good practice to start from an already existing code, to save it under another name, and to modify it according to your needs.

3. Modify the current code by adding all rows beginning with `Serial` as follows:

```
/*
   TalkingAndBlink250ms Program
   Turns a LED connected to digital pin 8 on for 250ms, then off
for 1s, infinitely.
   In both steps, the Arduino Board send data to the console of the
IDE for information purpose.
   Written by Julien Bayle, this example code is under Creative
Commons CC-BY-SA
 */

// Pin 8 is the one connected to our pretty LED
int ledPin = 8; // ledPin is an integer variable initialized at 8

// --------- setup routine
```

```
void setup() {
  pinMode(ledPin, OUTPUT); // initialize the digital pin as an
output
  Serial.begin(9600);      // Serial communication setup at 9600
baud
}// --------- loop routine
void loop() {
  digitalWrite(ledPin, HIGH);   // turn the LED on

  Serial.print("the pin ");     // print "the pin "
  Serial.print(ledPin);         // print ledPin's value (currently
8)
  Serial.println(" is on");     // print " is on"

  delay(250);                   // wait for 250ms in the current
state

  digitalWrite(ledPin, LOW);    // turn the LED off

  Serial.print("the pin ");     // print "the pin "
  Serial.print(ledPin);         // print ledPin's value (still 8)
  Serial.println(" is off");    // print " is off

  delay(1000);                  // wait for 1s in the current
state
}
```

Please notice that I highlight the comment code a bit each time in order to make things more readable. In the following steps, for instance, I won't write the following comment:

// ---------- loop routine

You can also find the whole code in the zip file in the folder Chapter02/TalkingAndBlink250ms/.

4. Click on the Serial Monitor button in the Arduino IDE:

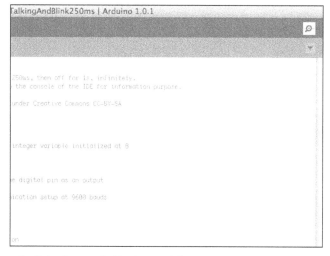

Click on the little glass symbol in the top-right corner to activate the Serial Monitor

5. Choose the same baud rate you wrote in the code, which is in the menu at the bottom-right of the Serial Monitoring window, and observe what is happening.

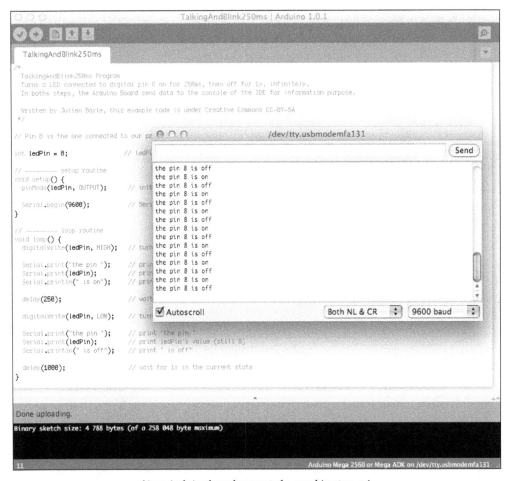

Your Arduino board seems to be speaking to you!

You will notice some messages appearing in the Serial Monitor window, synchronized with the blinking LED states.

Now, we can be sure that our code is fine because each message is sent and because all rows are processed sequentially; it means the `digitalWrite()` functions are also called correctly (nothing is blocked). This information can be a clue, for instance, to check our circuit once more to try to find the error there instead of in the code.

Of course this is a trivial example, but I'm sure you understand the target and the power of tracing your code!

Serial functions in more detail

Let's check what we added in the code.

Serial.begin()

Everything begins with the `Serial.begin()` function. This function in the `setup()` routine is executed only once, that is, when the Arduino is starting.

In the code, I set up the board to initiate a serial communication at 9,600 baud.

Serial.print() and Serial.println()

`Serial.print()` and `Serial.println()` behave almost identically: they write something to the serial output, but the `ln` version also adds a carriage return and a newline.

The syntax of this function is `Serial.print(val)` or `Serial.print(val,format)`.

You can pass one or two arguments.

Basically, if `Serial.print(5)` prints the number 5 as an ASCII-encoded decimal symbol, `Serial.print(5,OCT)` prints the number 5 as an ASCII-encoded octal one.

Digging a bit...

If you checked the code carefully (and I'm sure you did), you noticed we put two groups of three rows: one group just after the digitalWrite(ledPin,HIGH) function that lights on the LED, and the other group after the row that lights it off.

Got it?

We have asked the Arduino board to send a message according to the last order passed to the digital pin numbered 8, where the LED is still connected. And the board sends a message when we asked the pin to deliver current (when the LED is on), and another message when the pin doesn't deliver current (when the LED is off).

You just wrote your first trace routine.

Talking to the board from the computer

You probably noticed a text field and a **Send** button in the Serial Monitor window:

We can send symbol to our Arduino board using Serial Communication

This means we can also use that tool to send data to the board from our computer. The firmware's board, however, has to implement some other functions in order to be able to understand what we'd like to send.

Later in this book we'll see how to use the Serial Monitor window, the genius Processing framework, and the Max 6 framework to send messages easily to the Arduino board.

Summary

In this chapter, we learned about programming using C language. We also learned how to use the serial monitoring feature of our Arduino IDE in order to know a bit more about what is happening in real time in our Arduino processor using traces.

I spoke about serial communication because it is very useful and is also used in many real-life projects in which you need a computer and an Arduino board to communicate among themselves. It can also be used between two Arduino boards or between Arduino boards and other circuits.

In the next chapter, we'll enter C code by using the serial monitoring window in order to make things a bit less abstract.

3
C Basics – Making You Stronger

C programming isn't that hard. But it requires enough work at the beginning. Fortunately, I'm with you and we have a very good friend since three chapters – our Arduino board. We will now go deep into the C language, and I'll do my best to be more concrete and not abstract.

This chapter and the next one are truly C language-oriented because the Arduino program design requires knowledge in programming logic statements. After these two chapters, you'll be able to read any code in this book; these strong basics will also help you in further projects, even those not related to Arduino.

I will also progressively introduce new concepts that we will use later, such as functions. Don't be afraid if you don't understand it that well, I like my students to hear some words progressively sometimes even without a proper definition at first, because it helps further explanation.

So if I don't define it but talk about it, just relax, explanations are going to come further. Let's dive in.

Approaching variables and types of data

We already used variables in the previous chapters' examples. Now, let's understand this concept better.

What is a variable?

A **variable** is a memory storage location bounded to a symbolic name. This reserved memory area can be filled or left empty. Basically, it is used to store different types of values. We used the variable `ledPin` in our previous examples, with the keyword `int`.

Something very useful with variables is the fact that we can change their content (the value) at runtime; this is also why they are called variables, compared to constants that also store values, but that cannot be changed while the program is running.

What is a type?

Variables (and constants) are associated with a type. A type, also called **data type**, defines the possible nature of data. It also offers a nice way to directly reserve a space with a defined size in memory. C has around 10 main types of data that can be extended as we are going to see here.

I'm deliberately only explaining the types we'll use a lot in Arduino programming. This fits with approximately 80 percent of other usual C data types and will be more than enough here.

Basically, we are using a type when we declare a variable as shown here:

```
int ledPin; // declare a variable of the type int and named "ledPin"
```

A space of a particular size (the size related to the `int` type) is reserved in memory, and, as you can see, if you only write that line, there is still no data stored in that variable. But keep in mind that a memory space is reserved, ready to be used to store values.

Type	Definition	Size in memory
void	This particular type is used only in *function* declarations and while defining pointers with unknown types. We'll see that in the next chapter.	
boolean	It stores `false` or `true`.	1 byte (8 bit)
char	It stores single-quoted characters such as `'a'` as *numbers*, following the ASCII chart (`http://en.wikipedia.org/wiki/ASCII_chart`).	1 byte
	It is a *signed* type and stores numbers from -128 to 127; it can be unsigned and then stores numbers from 0 to 255.	

Type	Definition	Size in memory
byte	It stores numbers as *8-bit unsigned* data that means from 0 to 255.	8 bits
int	It stores numbers as *2-bytes signed* data which means from -32,768 to 32,767 it can also be unsigned and then store numbers from 0 to 65,535.	2 bytes (16 bit)
word	It stores numbers as *2-bytes unsigned* data exactly as *unsigned* int does.	2 bytes (16 bit)
long	It stores numbers as *4-bytes signed* data, which means from -2,147,483,648 to 2,147,483,647 and can be unsigned and then stores numbers from 0 to 4,294,967,295.	4 bytes (32 bit)
float	It basically stores numbers with a decimal point from -3.4028235E + 38 to 3.4028235E + 38 as *4-bytes signed* data. Be careful of the required precision; they only have six to seven decimal digits and can give strange rounding results sometimes.	4 bytes (32 bit)
double	It generally stores float values with a precision two times greater than the float value. Be careful, in the Arduino IDE and the board, the double implementation is exactly the same as float; that means with only six to seven decimal digits of precision.	4 bytes (32 bit)
Array	Array is an ordered structure of consecutive elements of the same type that can each be accessed with an index number.	number of elements x size of elements' type
string	It stores text strings in an array of char where the last element is null that is a particular character (ASCII code 0). Be careful of the "s" in lower case at the beginning of string.	number of elements * 1 byte
String	It is a particular structure of data, namely a class, that provides a nice way to use and work with strings of text. It comes with methods/functions to easily concatenate strings, split strings, and much more. Be careful of the capital "S" at the beginning of String.	available every time with the length() method

The roll over/wrap concept

If you go beyond the possible bounds of a type, the variable rolls over to the other side of the boundary.

The following is an example:

```
int myInt = 32767; //the maximum int value
myInt = myInt + 1; // myInt is now -32768
```

It happens in both directions, subtracting 1 from an int variable storing -32768 results in 32767. Keep that in mind.

Declaring and defining variables

We are going to describe how to declare then define variables and learn how to do both at the same time.

Declaring variables

Declaration of a variable is a statement in which you specify an *identifier*, a *type*, and eventually the variable's dimensions.

An identifier is what we call the *name of the variable*. You know what the type is too. The dimensions are useful for arrays, for instance, but also for String (which are processed as arrays internally).

In C and all other strongly-typed languages such as Java and Python, we *must* declare variables before using them. Anyway, the compiler will complain in case you forget the declaration.

Defining variables

The following table contains some examples of variable definition:

Type	Example

boolean	bool myVariable; // declaration of the variable
	myVariable = true; // definition of the variable by assigning it a value
	bool myOtherVariable = false; // declaration and definition inside the same statement !

char	char myChar = 'U'; // declaration and definition using the ASCII value of 'U' (i.e 85)
	char myOtherChar = 85; // equals the previous statement
	
	char myDefaultChar = 128; // this gives an ERROR because char are signed from -128 to 127
	unsigned char myUnsignedChar = 128; // this is correct !

byte	byte myByte = B10111; // 23 in binary notation with the B notation byte myOtherByte = 23; // equals the previous statement

int	int ledPin = 8; // classic for us, now :)
	unsigned myUint = 32768; // very okay with the prefix unsigned !

word	word myWord = 12345;

long	long myLong = -123; // don't forget that we can use negative numbers too!
	long myOtherLong = 345;
	unsigned myUlong = 2147483648; // correct because of the unsigned prefix

float	float myFloat = -123456.1; // they can be negative.
	float myOtherFloat = 1.234567; //
	float myNoDecimalPointedFloat = 1234; // they can have a decimal part equaling zero

double	double myDouble = 1.234567; // Arduino implementation of double is same as float

Type	Example

Array	`int myIntTable[5]; // declaration of a table that can contain 5 integers`
	`boolean myOtherTab[] = { false, true, true}; // declaration and definition of a 3 boolean arrays`
	
	`myIntTable[5]; // considering the previous definition, this gives an array bound ERROR (index starts from 0 and thus the last one here is myIntTable[4])`
	`myOtherTab[1]; // this elements can be manipulated as a boolean, it IS a boolean with the value true`

string	`char mystring[3]; // a string of 3 characters`
	`char mystring2[4] = {'b','y','t','e'}; // declaration & definition`
	`char mystring3[4] = "byte"; // equals to mystring2;`
	`char mystring4[ ] = "byte"; // equals to mystring3;`

Defining a variable is the act of assigning a value to the memory area previously reserved for that variable.

Let's declare and define some variables of each type. I have put some explanations in the code's comments.

Here are some examples you can use, but you'll see in each piece of code given in this book that different types of declaration and definition are used. You'll be okay with that as soon as we'll wire the board.

Let's dig a bit more into the `String` type.

String

The `String` type deserves a entire subchapter because it is a bit more than a type. Indeed, it is an object (in the sense of object-oriented programming).

Objects come with special properties and functions. Properties and functions are available natively because `String` is now a part of the Arduino core and can be seen as a pre-existing entity even if your IDE contains no line.

Again, the framework takes care of things for you, providing you a type/object with powerful and already coded functions that are directly usable.

Check out `http://arduino.cc/en/Reference/StringObject` in the Arduino website.

String definition is a construction

We talked about definition for variables, but objects have a similar concept called **construction**.

For String objects, I'm talking about *construction* instead of *definition* here but you can consider both terms equal. Declaring a String type in Arduino core involves an object constructor, which is an object-oriented programming concept; we don't have to handle it, fortunately, at this point.

```
String myString01 = "Hello my friend"; // usual constant string to
construct it
String myString02 = String('U'); // convert U char into a String
object

// concatenating 2 String together and put the result into another
String myString03 = String(myString01 + ", we are trying to play with
String(s));

// converting the current value of integer into a String object
int myNiceInt = 8; // define an integer
String myString04 = String(myNiceInt); // convert to a String object

// converting the current value of an integer w/ a base into a String
object
int myNiceInt = 47; // define an integer
String myString05 = String(myNiceInt, DEC);
String myString06 = String(myNiceInt, HEX);
String myString07 = String(myNiceInt, BIN);
```

Using indexes and search inside String

Strings are arrays of char elements. This means we can access any element of a String type through their indexes.

Keep in mind that indexes start at 0 and not at 1. The String objects implement some functions for this particular purpose.

charAt()

Considering a String type is declared and defined as follows:

```
String myString = "Hello World !!";
```

The statement `myString.charAt(3)` returns the fourth element of the string that is: 1. Notice the specific notation used here: we have the name of the `String` variable, a dot, then the name of the function (which is a method of the `String` object), and the parameter 3 which is passed to the function.

> The `charAt()` function returns a character at a particular position inside a **string**.
>
> **Syntax:** `string.charAt(int);`
>
> `int` is an integer representing an index of the `String` value.
>
> **Returns type:** `char`

Let's learn about other similar functions. You'll use them very often because, as we have already seen, communicating at a very low-level point of view includes parsing and processing data, which can very often be strings.

indexOf() and lastIndexOf()

Let's consider the same declaration/definition:

```
String myString = "Hello World !!";
```

`myString.indexOf('r')` equals 8. Indeed, r is at the ninth place of the value of the string `myString`. `indexOf(val)` and looks for the first occurrence of the value `val`.

If you want to begin your search from a particular point, you can specify a start point like that: `indexOf(val,start)`, where `start` is the index from where the function begins to search for the character `val` in the string. As you have probably understood, the second argument of this function (`start`) can be omitted, the search starts from the first element of the string by default, which is 0.

> The `indexOf()` function returns the first occurrence of a string or character inside a string.
>
> **Syntax:** `string.indexOf(val, from);`
>
> `val` is the value to search for which can be a string or a character. `from` is the index to start the search from, which is an `int` type. This argument can be omitted. The search goes forward.
>
> **Returns type:** `int`

Similarly, `lastIndexOf(val,start)` looks for the last occurrence of `val`, searching **backwards** from `start`, or from the last element if you omit `start`.

The `lastIndexOf()` function returns the last occurrence of a string or character inside a string.

> **Syntax**: `string.lastIndexOf(val, from);`
>
> `val` is the value to search for which is a string or a character. `from` is the index to start the search from which is an `int` type. This argument can be omitted. The search goes backwards.
>
> **Returns type**: `int`

startsWith() and endsWith()

The `startsWith()` and `endsWith()` functions check whether a string starts or ends with, respectively, another string passed as an argument to the function.

```
String myString = "Hello World !!";
String anotherString ="Hell" ;
myString.startsWith(anotherString); // this returns true
myString.startsWith("World"); // this returns false
```

> The `startsWith()` function returns `true` if a string starts with the same characters as another string.
>
> **Syntax**: `string.startsWith(string2);`
>
> `string2` is the string pattern with which you want to test the string.
>
> **Returns type**: `boolean`

I guess, you have begun to understand right now. `endsWith()` works like that too, but compares the string pattern with the end of the string tested.

> The `endsWith()` function returns `true` if a string ends with the same characters as another string.
>
> **Syntax**: `string.endsWith(string2);`
>
> `string2` is the string pattern with which you want to test the string.
>
> **Returns type**: `boolean`

Concatenation, extraction, and replacement

The preceding operations also introduce new C operators. I'm using them here with strings but you'll learn a bit more about them in a more global context further.

Concatenation

Concatenation of strings is an operation in which you take two strings and you glue them together. It results in a new string composed of the previous two strings. The order is important; you have to manage which string you want appended to the end of the other.

Concat()

Arduino core comes with the `string.concat()` function, which is especially designed for this purpose.

```
String firstString = "Hello ";
String secondString ="World!";


// appending the second to the first and put the result in the first
firstString.concat(secondString);
```

The `concat()` function appends one string to another (that is concatenate in a defined order).

Syntax: `string.concat(string2);`

`string2` is a string and is appended to the end of string. Remember that, the previous content of the string is overwritten as a result of the concatenation.

Returns type: `int` (the function returns 1 if the concatenation happens correctly).

Using the + operator on strings

There is another way to concatenate two strings. That one doesn't use a function but an operator: +.

```
String firstString = "Hello ";
String secondString ="World!";


// appending the second to the first and putting the result in the
first
firstString = firstString + secondString;
```

This code is the same as the previous one. + is an operator that I'll describe better a bit later. I'm giving you something more here: a condensed notation for the + operator:

```
firstString = firstString + secondString;
```

This can also be written as:

```
firstString += secondString;
```

Try it. You'll understand.

Extract and replace

String manipulation and alteration can be done using some very useful functions extracting and replacing elements in the string.

substring() is the extractor

You want to extract a part of a string. Imagine if the Arduino board sends messages with a specific and defined communication protocol:

```
<output number>.<value>
```

The output number is coded with two characters every time, and the value with three (45 has to be written as 045). I often work like that and pop out these kind of messages from the serial port of my computer via the USB when I need to; for instance, send a command to light up a particular LED with a particular intensity. If I want to light the LED on the fourth output at 100/127, I send:

```
04.100
```

Arduino *needs* to understand this message. Without going further with the communication protocol design, as that will be covered in *Chapter 7, Talking Over Serial*, I want to introduce you to a new feature — splitting strings.

```
String receivedMessage = "04.100";
String currentOutputNumber;
String currentValueNumber;

// extracting a part of receivedMessage from index 0 (included) to 1
(excluded)
currentOutputNumber = receivedMessage.substring(0,2);

// extracting a part of receivedMessage from index 3 (included) to the
end
currentValueNumber = receivedMessage.substring(3);
```

This piece of code splits the message received by Arduino into two parts.

The substring() function extracts a part of a string from a start index (included) to another (not included).

Syntax: string.substring(from, to);

from is the start index. The result includes the content of the from string element. to is the end index. The result doesn't include the content of the end string element, it can be omitted.

Returns type: String

Let's push the concept of string extract and split it a bit further.

Splitting a string using a separator

Let's challenge ourselves a bit. Imagine I don't know or I'm not sure about the message format (two characters, a dot, and three characters, that we have just seen). This is a real life case; while learning to make things, we often meet strange cases where those *things* don't behave as expected.

Imagine I want to use the dot as a separator, because I'm very sure about it. How can I do that using the things that we have already learned? I'd need to extract characters. OK, I know substring() now!

But I also need an index to extract the content at a particular place. I also know how to find the index of an occurrence of a character in a string, using indexOf().

Here is how we do that:

```
String receivedMessage = "04.100";
String currentOutputNumber;
String currentValueNumber;
int splitPointIndex;

// storing the index of the separator in the String
splitPointIndex = receivedMessage.indexOf('.');

// extracting my two elements
currentOutputNumber = receivedMessage.substring(0, splitPointIndex);
currentValueNumber = receivedMessage.substring(splitPointIndex + 1);
```

Firstly, I find the split point index (the place in the string where the dot sits). Secondly, I use this result as the last element of my extracted substring. Don't worry, the last element isn't included, which means currentOutputNumber doesn't contain the dot.

At last, I'm using `splitPointIndex` one more time as the start of the second part of the string that I need to extract. And what? I add the integer `1` to it because, as you master `substring()` now and know, the element corresponding to the start index is always included by the `substring()` operation. We don't want that dot because it is only a separator. Right?

Don't worry if you are a bit lost. Things will become clearer in the next subchapters and especially when we'll make Arduino process things, which will come a bit later in the book.

Replacement

Replacements are often used when we want to convert a communication protocol to another. For instance, we need to replace a part of a string by another to prepare a further process.

Let's take our previous example. We now want to replace the dot by another character because we want to send the result to another process that only understands the space character as a separator.

```
String receivedMessage = "04.100";
String originalMessage;

// keeping a trace of the previous message by putting it into another
variable
originalMessage = receivedMessage;

// replacing dot by space character in receivedMessage
receivedMessage.replace('.',' ');
```

Firstly, I put the content of the `receivedMessage` variable into another variable named `originalMessage` because I know the `replace()` function will definitely modify the processed string. Then I process `receivedMessage` with the `replace()` function.

The `replace()` function replaces a part of a string with another string.

Syntax: `string.replace(substringToReplace, replacingSubstring);`

`from` is the start index. The result includes the content of a `from` string element. `to` is the end index. The result doesn't include the content of an end string element, it can be omitted. Remember that, the previous content of the string is overwritten as a result of the replacement (copy it to another string variable if you want to keep it).

Returns type: `int` (the function returns 1 if the concatenation happens correctly).

This function can, obviously, replace a character by another character of course.
A string is an array of characters. It is not strange that one character can be processed as a string with only one element. Let's think about it a bit.

Other string functions

There are some other string processing functions I'd like to quickly quote here.

toCharArray()

This function copies all the string's characters into a "real" character array, also named, for internal reasons, a buffer. You can check `http://arduino.cc/en/Reference/StringToCharArray`.

toLowerCase() and toUpperCase()

These functions replace the strings processed by them by the same string but with all characters in lowercase and uppercase respectively. You can check `http://arduino.cc/en/Reference/StringToLower` and `http://arduino.cc/en/Reference/StringToUpperCase`. Be careful, as it overwrites the string processed with the result of this process.

trim()

This function removes all whitespace in your string. You can check `http://arduino.cc/en/Reference/StringTrim`. Again, be careful, as it overwrites the strings processed with the result of this process.

length()

I wanted to end with this functioin. This is the one you'll use a lot. It provides the length of a string as an integer. You can check `http://arduino.cc/en/Reference/StringLength`.

Testing variables on the board

The following is a piece of code that you can also find in the folder `Chapter03/VariablesVariations/`:

```
/*
  Variables Variations Program
 This firmware pops out messages over Serial to better understand
 variables' use.
```

```
Switch on the Serial Monitoring window and reset the board after
that.
 Observe and check the code :)

 Written by Julien Bayle, this example code is under Creative Commons
CC-BY-SA
 */



// declaring variables before having fun !
boolean myBoolean;
char myChar;
int myInt;
float myFloat;
String myString;


void setup(){
  Serial.begin(9600);
  myBoolean = false;
  myChar = 'A';
  myInt = 1;
  myFloat = 5.6789 ;
  myString = "Hello Human!!";
}

void loop(){

  // checking the boolean
  if (myBoolean) {
    Serial.println("myBoolean is true");
  }
  else {
    Serial.println("myBoolean is false");
  }


  // playing with char & int
  Serial.print("myChar is currently ");
  Serial.write(myChar);
  Serial.println();

  Serial.print("myInt is currently ");
  Serial.print(myInt);
```

```
    Serial.println();

    Serial.print("Then, here is myChar + myInt : ");
    Serial.write(myChar + myInt);
    Serial.println();


    // playing with float & int
    Serial.print("myFloat is : ");
    Serial.print(myFloat);
    Serial.println();

    // putting the content of myFloat into myInt
    myInt = myFloat;
    Serial.print("I put myFloat into myInt, and here is myInt now : ");
    Serial.println(myInt);


    // playing with String
    Serial.print("myString is currently: ");
    Serial.println(myString);

    myString += myChar; // concatening myString with myChar
    Serial.print("myString has a length of ");
    Serial.print(myString.length());// printing the myString length
    Serial.print(" and equals now: ");
    Serial.println(myString);

  // myString becomes too long, more than 15, removing the last 3
elements
  if (myString.length() >= 15){
    Serial.println("myString too long ... come on, let's clean it up!
");
    myInt = myString.lastIndexOf('!'); // finding the place of the '!'
    myString = myString.substring(0,myInt+1);  // removing characters

    Serial.print("myString is now cleaner: ");
    Serial.println(myString);

    // putting true into myBoolean
  }
  else {
    myBoolean = false;     // resetting myBoolean to false
  }


    delay(5000);     // let's make a pause
```

```
    // let's put 2 blank lines to have a clear read
    Serial.println();
    Serial.println();
}
```

Upload this code to your board, then switch on the serial monitor. At last, reset the board by pushing the reset button and observe. The board writes directly to your serial monitor as shown in the following screenshot:

The serial monitor showing you what your board is saying

Some explanations

All explanations will come progressively, but here is a small summary of what is happening right now.

I first declare my variables and then define some in `setup()`. I could have declared and defined them at the same time.

Refreshing your memory, `setup()` is executed only one time at the board startup. Then, the `loop()` function is executed infinitely, sequentially running each row of statement.

In `loop()`, I'm first testing `myBoolean`, introducing the `if()` conditional statement. We'll learn this in this chapter too.

Then, I'll play a bit with the `char`, `int`, and `String` types, printing some variables, then modifying them and reprinting them.

The main point to note here is the `if()` and `else` structure. Look at it, then relax, answers will come very soon.

The scope concept

The scope can be defined as a particular property of a variable (and functions, as we'll see further). Considering the source code, the scope of a variable is that part of the code where this variable is visible and usable.

A variable can be *global* and then is visible and usable everywhere in the source code. But a variable can also be *local*, declared inside a function, for instance, and that is visible only inside this particular function.

The scope property is *implicitly* set by the place of the variable's declaration in the code. You probably just understood that every variable could be declared globally. Usually, I follow my own *digital haiku*.

 Let each part of your code know only variables that it has to know, no more.

Trying to minimize the scope of the variables is definitely a winning way. Check out the following example:

```
// this variable is declared at the highest level, making it visible
everywhere
int globalString;

void setup(){
  // … some code
}
void loop(){
   int a; // a is visible inside the loop function only
   anotherFunction(); // calling the global function anotherFunction

// … some other code
}

void anotherFunction() {
  // … yet another code
   int veryLocalVar; // veryLocalVar is visible only in anotherFunction
function
}
```

We could represent the code's scope as a box more or less imbricated.

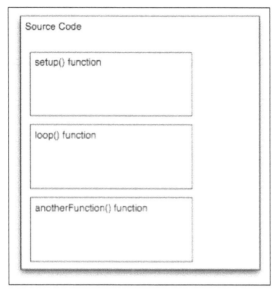

Code's scope seen as boxes

The external box represents the source code's highest level of scope. Everything declared at this level is visible and usable by all functions; it is the global level.

Every other box represents a particular scope in itself. Every variable declared in one scope cannot be seen and used in higher scopes neither in the same level ones.

This representation is very useful to my students who always need more visuals. We'll also use this metaphor while we talk about *libraries*, especially. What is declared in libraries can be used in our code if we include some specific headers at the beginning of the code, of course.

static, volatile, and const qualifiers

Qualifiers are the keywords that are used to change the processor's behavior considering the *qualified* variable. In reality, the compiler will use these qualifiers to change characteristics of the considered variables in the binary firmware produced. We are going to learn about three qualifiers: `static`, `volatile`, and `const`.

static

When you use the `static` qualifier for a variable inside a function, this makes the variable persistent between two calls of the function. Declaring a variable inside a function makes the variable, implicitly, local to the function as we just learned. It means only the function can know and use the variable. For instance:

```
int myGlobalVariable;

void setup(){
}

void loop(){
  myFunction(digitalPinValue);
}

void myFunction(argument){
int aLocalVariable;
aLocalVariable = aLocalVariable + argument;
  // playing with aLocalVariable
}
```

This variable is seen in the `myFunction` function only. But what happens after the first loop? The previous value is lost and as soon as `int aLocalVariable;` is executed, a new variable is set up, with a value of zero. Check out this new piece of code:

```
int myGlobalVariable;
void setup(){
}

void loop(){
  myFunction(digitalPinValue);
}

void myFunction(argument){
static int aStaticVariable;
aStaticVariable = aStaticVariable + argument;
  // playing with aStaticVariable
}
```

This variable is seen in the `myFunction` function only and, after adding an argument has modified it, we can play with its new value.

In this case, the variable is qualified as `static`. It means the variable is declared *only* the first time. This provides a useful way to keep trace of something and, at the same time, make the variable, containing this trace, local.

volatile

When you use the `volatile` qualifier in a variable declaration statement, this variable is loaded from the RAM instead of the storage register memory space of the board. The difference is subtle and this qualifier is used in specific cases where your code itself doesn't have the control of something else executed on the processor. One example, among others, is the use of interrupts. We'll see that a bit later.

Basically, your code runs normally, and some instructions are triggered not by this code, but by another process such as an external event. Indeed, our code doesn't know when and what **Interrupt Service Routine (ISR)** does, but it stops when something like that occurs, letting the CPU run ISR, then it continues. Loading the variable from the RAM prevents some possible inconsistencies of variable value.

const

The `const` qualifier means constant. Qualifying a variable with `const` makes it unvariable, which can sound weird.

If you try to write a value to a `const` variable after its declaration/definition statement, the compiler gives an error. The scope's concept applies here too; we can qualify a variable declared inside a function, or globally. This statement defines and declares the `masterMidiChannel` variable as a constant:

```
const int masterMidiChannel = 10;
```

This is equivalent to:

```
#define masterMidiChannel 10
```

 There is *no* semicolon after a `#define` statement.

`#define` seems a bit less used as `const`, probably because it cannot be used for constant arrays. Whatever the case, `const` can always be used. Now, let's move on and learn some new operators.

Operators, operator structures, and precedence

We have already met a lot of operators. Let's first check the arithmetic operators.

Arithmetic operators and types

Arithmetic operators are:

- + (plus sign)
- - (minus)
- * (asterisk)
- / (slash)
- % (percent)
- = (equal)

I'm beginning with the last one: =. It is the **assignment** operator. We have already used it a lot to define a variable, which just means to assign a value to it. For instance:

```
int oscillatorFrequency = 440;
```

For the other operators, I'm going to distinguish two different cases in the following: character types, which include char and String, and numerical types. Operators can change their effect a bit according to the types of variables.

Character types

char and String can only be processed by +. As you may have guessed, + is the concatenation operator:

```
String myString = "Hello ";
String myString2 = "World";

String myResultString = myString + myString2;
myString.concat(myString2);
```

In this code, concatenation of myResultString and myString results in the Hello World string.

Numerical types

With all numerical types (int, word, long, float, double), you can use the following operators:

- + (addition)
- - (subtraction)
- * (multiplication)
- / (division)
- % (modulo)

A basic example of multiplication is shown as follows:

```
float OutputOscillatorAmplitude = 5.5;
int multiplier = 3;
OutputOscillatorAmplitude = OutputOscillatorAmplitude * multiplier
```

 As soon as you use a float or double type as one of the operand, the floating point calculation process will be used.

In the previous code, the result of OutputOscillatorAmplitude * multiplier is a float value. Of course, division by zero is *prohibited*; the reason is math instead of C or Arduino.

Modulo is simply the remainder of the division of one integer by another one. We'll use it a lot to keep variables into a controlled and chosen range. If you make a variable grow to infinite but manipulate its modulo by 7 for instance, the result will always be between 0 (when the growing variable will be a multiple of 7) and 6, constraining the growing variable.

Condensed notations and precedence

As you may have noticed, there is a condensed way of writing an operation with these previously explained operators. Let's see two equivalent notations and explain this.

Example 1:

```
int myInt1 = 1;
int myInt2 = 2;

myInt1 = myInt1 + myInt2;
```

Example 2:

```
int myInt1 = 1;
int myInt2 = 2;

myInt1 += myInt2;
```

These two pieces of code are equivalent. The first one teaches you about the precedence of operators. There is a table given in *Appendix B, Operator Precedence in C and C++* with all precedencies. Let's learn some right now.

+, -, *, /, and % have a greater precedence over =. That means `myInt1 + myInt2` is calculated before the assignment operator, then, the result is assigned to `myInt1`.

The second one is the condensed version. It is equivalent to the first version and thus, precedence applies here too. A little tricky example is shown as follows:

```
int myInt1 = 1;
int myInt2 = 2;

myInt1 += myInt2 + myInt2;
```

You need to know that + has a higher precedence over +=. It means the order of operations is: first, `myInt2 + myInt2` then `myInt1` + the result of the freshly made calculation `myInt2 + myInt2`. Then, the result of the second is assigned to `myInt1`. This means it is equivalent to:

```
int myInt1 = 1;
int myInt2 = 2;

myInt1 = myInt1 + myInt2 + myInt2;
```

Increment and decrement operators

I want to point you to another condensed notation you'll meet often: the double operator.

```
int myInt1 = 1;

myInt++;   // myInt1 now contains 2
myInt--;   // myInt1 now contains 1
```

++ is equivalent to +=1, -- is equivalent to -=1. These are called *suffix increment* (++) and *suffix decrement* (--). They can also be used as *prefix*. ++ and -- as prefixes have lower precedencies than their equivalent used as suffix but in both cases, the precedence is very much higher than +, -, /, *, and even = and +=.

The following is a condensed table I can give you with the most used cases. In each group, the operators have the same precedence. It drives the expression `myInt++ + 3` to be ambiguous. Here, the use of parenthesis helps to define which calculation will be made first.

Precedencies groups	Operators	Names
2	++	Suffix increment
	--	Suffix decrement
	()	Function call
	[]	Array element access
3	++	Prefix increment
	--	Prefix decrement
5	*	Multiplication
	/	Division
	%	Modulo
6	+	Addition
	-	Subtraction
16	=	Assignment
	+=	Assignment by sum
	-=	Assignment by difference
	*=	Assignment by product
	/=	Assignment by quotient
	%=	Assignment by remainder

I guess you begin to feel a bit better with operators, right? Let's continue with a very important step: types conversion.

Types manipulations

When you design a program, there is an important step consisting of choosing the right type for each variable.

Choosing the right type

Sometimes, the choice is constrained by external factors. This happens when, for instance, you use the Arduino with an external sensor able to send data coded as integers in 10 bits (2^{10} = 1024 steps of resolution). Would you choose `byte` type knowing it only provides a way to store number from 0 to 255? Probably not! You'll choose `int`.

Sometimes you have to choose it yourself. Imagine you have data coming to the board from a Max 6 framework patch on the computer via your serial connection (using USB). Because it is the most convenient, since you designed it like that, the patch pops out `float` numbers encapsulated into string messages to the board. After having parsed, cut those messages into pieces to extract the information you need (which is the `float` part), would you choose to store it into `int`?

That one is a bit more difficult to answer. It involves a *conversion* process.

Implicit and explicit types conversions

Type conversion is the process that changes an entity data type into another. Please notice I didn't talk about variable, but entity.

It is a consequence of C design that we can convert only the values stored in variables, others keep their type until their lives end, which is when the program's execution ends.

Type conversion can be *implicitly* done or *explicitly* made. To be sure everyone is with me here, I'll state that *implicitly means not visibly and consciously written*, compared to *explicitly that means specifically written in code*, here.

Implicit type conversion

Sometimes, it is also called *coercion*. This happens when you don't specify anything for the compiler that has to make an automatic conversion following its own basic (but often smart enough) rules. The classic example is the conversion of a `float` value into an `int` value.

```
float myFloat = 12345.6789 ;
int myInt;
myInt = myFloat;

println(myInt); // displays 12345
```

I'm using the assignment operator (=) to put the content of `myFloat` into `myInt`. It causes **truncation** of the `float` value, that is, the *removal of the decimal part*. You have definitely lost something if you continue to work only with the `myInt` variable instead of `myFloat`. It can be okay, but you have to keep it in mind.

Another less classic example is the implicit conversion of `int` type to `float`. `int` doesn't have a decimal part. The implicit conversion to `float` won't produce something other than a decimal part that equals zero. This is the easy part.

But be careful, you could be surprised by the implicit conversion of int to float. Integers are encoded over 32 bits, but float, even if they are 32 bits, have a *significand* (also called mantissa) encoded over 23 bits. If you don't remember this precisely, it is okay. But I want you to remember this example:

```
float myFloat;
long int myInt = 123456789;
void setup(){
  Serial.begin(9600);
  myFloat = myInt;
}

void loop(){
  Serial.println(myFloat); // displays a very strange result
}
```

The output of the code is shown as follows:

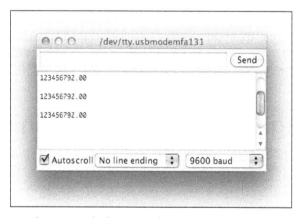

Strange results from int to float implicit conversion

I stored 123456789 into a long int type, which is totally legal (long int are 32-bits signed integers that are able to store integers from -2147483648 to 2147483647). After the assignment, I'm displaying the result that is: **123456792.00**. We expected 123456789.00 of course.

Implicit types conversions rules:

- long int to float can cause wrong results
- float to int removes the decimal part
- double to float rounds digit of double
- long int to int drops the encoded higher bits

Explicit type conversion

If you want to have predictable results, every time you can convert types explicitly. There are six conversion functions included in the Arduino core:

- char()
- int()
- float()
- word()
- byte()
- long()

We can use them by passing the variable you want to convert as an argument of the function. For instance, `myFloat = float(myInt);` where `myFloat` is a `float` type and `myInt` is an `int` type. Don't worry about the use, we'll use them a bit later in our firmware.

My rule about conversion: Take care of each type conversion you make. None should be obvious for you and it can cause an error in your logic, even if the syntax is totally correct.

Comparing values and Boolean operators

We now know how to store entities into variables, convert values, and choose the right conversion method. We are now going to learn how to compare variable values.

Comparison expressions

There are six comparison operators:

- `==` (equal)
- `!=` (not equal)
- `<` (less than)
- `>` (greater than)
- `<=` (less than or equal to)
- `>=` (greater than or equal to)

The following is a comparison expression in code:

```
int myInt1 = 4;
float myFloat = 5.76;
(myInt1 > myFloat) ;
```

An expression like that does nothing, but it is legal. Comparing two elements produces a result and in this small example, it isn't used to trigger or make anything. `myInt1 > myFloat` is a comparison expression. The result is, obviously, `true` or `false`, I mean it is a `boolean` value. Here it is `false` because 4 is not greater than 5.76. We can also combine comparison expressions together to create more complex expressions.

Combining comparisons with Boolean operators

There are three Boolean operators:

- `&&` (and)
- `||` (or)
- `!` (not)

It is time to remember some logic operations using three small tables. You can read those tables like column element + comparison operator + row element; the result of the operation is at the intersection of the column and the row.

The binary operator AND, also written as `&&`:

`&&`	true	false
true	true	false
false	false	false

Then the binary operator OR, also written as `||`:

`||`	true	false
true	true	true
false	true	false

Lastly, the unary operator NOT, also written as `!`:

	true	false
`!`	false	true

For instance, true && false = false, false || true = true. && and || are *binary operators*, they can *compare* two expressions.

! is a *unary operator* and can only work with one expression, negating it logically. && is the logical AND. It is true when both expressions compared are true, false in all other cases. || is the logic OR. It is true when one expression at least is true, false when they are both false. It is the inclusive OR. ! is the negation operator, the NOT. It basically inverts false and true into true and false.

These different operations are really useful and necessary when you want to carry out some tests in your code. For instance, if you want to compare a variable to a specific value.

Combining negation and comparisons

Considering two expressions A and B:

- NOT(A && B) = (NOT A || NOT B)
- NOT (A || B) = (NOT A && NOT B)

This can be more than useful when you'll create conditions in your code. For instance, let's consider two expressions with four variables a, b, c, and d:

- a < b
- c >= d

What is the meaning of !(a < b)? It is the negation of the expression, where:

!(a < b) equals (a >= b)

The opposite of *a strictly smaller than b* is *a greater than or equal to b*. In the same way:

!(c >= d) equals (c < d)

Now, let's combine a bit. Let's negate the global expression:

(a < b) && (c >= d) and !((a < b) && (c >= d)) equals (!(a < b) || !(c >= d)) equals (a >= b) || (c < d)

Here is another example of combination introducing the *operators precedence* concept:

```
int myInt1 = 4;
float myFloat = 5.76;
int myInt2 = 16;
(myInt1 > myFloat && myInt2 < myFloat) ;
( (myInt1 > myFloat) && (myInt2 < myFloat) ) ;
```

Both of my statements are equivalent. Precedence occurs here and we can now add these operators to the previous precedencies table (check *Appendix B, Operator Precedence in C and C++*). I'm adding the comparison operator:

Precedencies groups	Operators	Names
2	++	Suffix increment
	--	Suffix decrement
	( )	Function call
	[ ]	Array element access
3	++	Prefix increment
	--	Prefix decrement
5	*	Multiplication
	/	Division
	%	Modulo
6	+	Addition
	-	Subtraction
8	<	Less than
	<=	Less than or equal to
	>	Greater than
	>=	Greater than or equal to
9	==	Equal to
	!=	Not equal to
13	&&	Logical AND
14	||	Logical OR
15	? :	Ternary conditional
16	=	Assignment
	+=	Assignment by sum
	-=	Assignment by difference
	*=	Assignment by product
	/=	Assignment by quotient
	%=	Assignment by remainder

As usual, I cheated a bit and added the precedence group 15 that contains a unique operator, the ternary conditional operator that we will see a bit later. Let's move to conditional structures.

Adding conditions in the code

Because I studied Biology and have a Master's diploma, I'm familiar with organic and living behaviors. I like to tell my students that the code, especially in interaction design fields of work, has to be alive. With Arduino, we often build machines that are able to "feel" the real world and interact with it by *acting* on it. This couldn't be done without *condition* statements. This type of statement is called a control structure. We used one conditional structure while we tested our big code including variables display and more.

if and else conditional structure

This is the one we used without explaining. You just learned patience and zen. Things begin to come, right? Now, let's explain it. This structure is very intuitive because it is very similar to any conditional pseudo code. Here is one:

If the value of the variable a is smaller than the value of variable b, switch on the LED. Else switch it off.

Now the real C code, where I simplify the part about the LED by giving a state of 1 or 0 depending on what I want to further do in the code:

```
int a;
int b;
int ledState;
if (a < b) {
ledState = 1;
}
else {
ledState = 0;
}
```

I guess it is clear enough. Here is the general syntax of this structure:

```
If (expression) {
// code executed only if expression is true
}
else {
// code executed only if expression is false
}
```

Expression evaluation generally results in a Boolean value. But numerical values as a result of an expression in this structure can be okay too, even if a bit less explicit, which I, personally, don't like. An expression resulting in the numerical value 0 equals false in C in Arduino core, and equals true for any other values.

Being implicit often means making your code shorter and cuter. In my humble opinion, it also means to be very unhappy when you have to support and maintain a code several months later when it includes a bunch of implicit things without any comments.

I push my students to be explicit and verbose. We are not here to code things to a too small amount of memory, believe me. We are not talking about reducing a 3 megabytes code to 500 kilobytes but more about reducing a 200 kilobytes code to 198 kilobytes.

Chaining an if…else structure to another if…else structure

The following is the modified example:

```
int a;
int b;
int ledState;
if (a < b) {
ledState = 1;
}
else if (b > 0) {
ledState = 0;
}
else {
ledState = 1;
}
```

The first `if` test is: if a is smaller than b. If it is `true`, we put the value 1 inside the variable `ledState`. If it is `false`, we go to the next statement `else`.

This `else` contains another test on b: is b greater than 0? If it is, we put the value 0 inside the variable `ledState`. If it is `false`, we can go to the last case, the last `else`, and we put the value 1 inside the variable `ledState`.

One frequent error – missing some cases

Sometimes, the `if… else` chain is so complicated and long that we may miss some case and no case is verified. Be clear and try to check the whole universe of cases and to code the conditions according to it.

A nice tip is to try to put all cases on paper and try to find the *holes*. I mean, where the part of the variable values are not matched by tests.

if...else structure with combined comparisons expressions

The following is the previous example where I commented a bit more:

```
int a;
int b;
int ledState;
if (a < b) {      // a < b

ledState = 1;
}
else if (b > 0) {   // a >= b and b > 0
ledState = 0;
}
else {        // a >= b and b < 0
ledState = 1;
}
```

We can also write it in the following way considering the comment I wrote previously in the code:

```
int a;
int b;
int ledState;
if (a < b || (a >= b && b < 0) ) {
ledState = 1;
}
else if (a >= b && b > 0) {
ledState = 0;
}
```

It could be considered as a more condensed version where you have all statements for the switch on the LED in one place, same for switching it off.

Finding all cases for a conditional structure

Suppose you want to test a temperature value. You have two specific limits/points at which you want the Arduino to react and, for instance, alert you by lighting an LED or whatever event to interact with the real world. For instance, the two limits are: 15-degree Celsius and 30-degree Celsius. How to be sure I have all my cases? The best way is to use a pen, a paper, and to draw a bit.

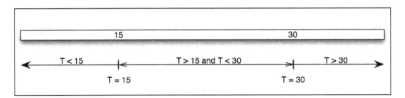

Checking all possible T values

We have three parts:

- T < 15
- T > 15 but T <30
- T > 30

So we have three cases:

- T< 15
- T >15 and T < 30
- T > 30

What happens when T = 30 or T = 15? These are holes in our logic. Depending on how we designed our code, it could happen. Matching all cases would mean: include T = 15 and T = 30 cases too. We can do that as follows:

```
float T ; // my temperature variable

if (T < 15) {
colorizeLed(blue);
}
else if (T >= 15 && T < 30) {
colorizeLed(white);
}
else if (T >= 30) {
colorizeLed(red);
}
```

I included these two cases into my comparisons. 15-degree Celsius is included in the second temperature interval and 30-degree Celsius in the last one. This is an example of how we can do it.

I'd like you to remember to use a pen and a paper in this kind of cases. This will help you to design and especially make some breaks from the IDE that is, in designing steps, really good. Let's now explore a new conditional structure.

switch...case...break conditional structure

Here, we are going to see a new conditional structure. The standard syntax is shown as follows:

```
switch (var) {
  case label:
  // statements
  break;
```

```
    case label:
    // statements
    break;
    default:
    // statements
}
```

var is compared for equality to each case label. If var equals a particular label value, the statements in this case are executed until the next break. If there is no match and you have used the optional default: case, the statements of this case are executed. Without the default: case, nothing is done. label must be a value, not a character, not string. Let's take a more concrete example:

```
float midiCCMessage;
switch (midiCCMessage) {
    case 7:
    changeVolume();
    break;
    case 10:
    changePan();
    break;
    default:
    LedOff();
}
```

This code is equivalent to:

```
float midiCCMessage;
if (midiCCMessage == 7) changeVolume();
else if (midiCCMessage == 10) changePan ();
else LedOff();
```

Are you okay?

 What I want to say is, when you want to compare a variable to many unique values, use switch...case... break, else use if...else.

When you have comparison intervals, if...else is more convenient because you can use < and > whereas in switch...case...break you cannot. Of course, we could combine both. But remember to keep your code as simple as you can.

Ternary operator

This strange notation is often totally unknown to my students. I used to say, "Hey! This is more C than Arduino" when they answer "That is why we have forgotten about it". Naughty students! This ternary operator takes three elements as input. The syntax is `(expression) ? val1 : val2`.

The expression is tested. If it is `true`, this whole statement returns (or equals) `val1`, if it is `false`, it equals `val2`.

Again imagine our Arduino, the temperature sensor, and only one limit which is 20 degree Celsius. I want to turn the LED blue if `T` is smaller than the limit, and red if `T` is greater or equal to 20 degree Celsius. Here is how we would use the two ternary operators:

```
Int T;
Int ledColor; // 0 means blue, 1 means red
ledColor = (T < 20) ? 0 : 1;
```

It can be a nice notation, especially if you don't need statement execution in each case but only variable assignments.

Making smart loops for repetitive tasks

A **loop** is a series of events repeating themselves in time. Basically, computers have been designed, at first, to make a lot of calculations repeatedly to save human's time. Designing a loop to repeat tasks that have to be repeated seems a natural idea. C natively implements some ways to design loops. Arduino core naturally includes three loop structures:

- `for`
- `while`
- `do...while`

for loop structure

The `for` loop statement is quite easy to use. It is based on, at least, one counter starting from a particular value you define, and increments or decrements it until another defined value. The syntax is:

```
for (declaration & definition ; condition ; increment) {
// statements
}
```

The counter is also named `index`. I'm showing you a real example here:

```
for (int i = 0 ; i < 100 ; i++) {
println(i);
}
```

This basic example defines a loop that prints all integers from 0 to 99. The declaration/definition of the integer type variable i is the first element of the `for` structure. Then, the condition describes in which case the statements included in this loop have to be executed. At last, the i++ increment occurs.

Pay attention to the increment element. It is defined with the increment as a suffix. It means here that the increment occurs after the end of the execution of the statements for a considered i value.

Let's break the loop for the first two and last two i values and see what happens. Declaration of the integer variable i for the first and second iteration is shown as follows:

- i = 0, is i smaller than 100? yes, `println(0)`, increment i
- i = 1, is i smaller than 100? yes, `println(1)`, increment i

For the last two iterations the value of i is shown as follows:

- i = 99, is i smaller than 100? yes, `println(99)`, increment i
- i = 100, is i smaller than 100? no, stop the loop

Of course, the index could be declared before the `for` structure, and only defined inside the `for` structure. We could also have declared and defined the variable before and we would have:

```
int i = 0;
for ( ; i < 100 ; i++) {
println(i);
}
```

This seems a bit strange, but totally legal in C and for the Arduino core too.

The scope of index

If the index has been declared inside the `for` loop parenthesis, its scope is only the `for` loop. This means that this variable is *not* known or *not* usable outside of the loop.

It normally works like that for any variable declared inside the statements of a `for` loop. This isn't something to do, even if it is totally legal in C. Why not? Because it would mean you'd declare a variable each time the loop runs, which isn't really smart. It is better to declare it outside of the loop, one time, then to use it inside of it, whatever the purpose (index, or variable to work with inside statements).

Playing with increment

Increment can be something more complex than only using the increment operator.

More complex increments

First, instead of writing `i++`, we could have written `i = i + 1`. We can also use other kind of operations like subtraction, multiplication, division, modulo, or combinations. Imagine that you want to print only odd numbers. Odd numbers are all of the form 2n + 1 where *n* is an integer. Here is the code to print odd numbers from 1 to 99:

```
for (int i = 0 ; i < 50 ; i = 2 * i + 1) {
println(i);
}
```

First values of `i` are: 1, 3, 5, and so on.

Decrements are negative increments

I just want to remix the previous code into something else in order to shake your mind a bit around increments and decrements. Here is another code making the same thing but printing odd numbers from 99 to 1:

```
for (int i = 50 ; i > 0 ; i = 2 * i - 1) {
println(i);
}
```

All right? Let's complicate things a bit.

Using imbricated for loops or two indexes

It is also possible to use more than one index in a `for` structure. Imagine we want to calculate a multiplication table until 10 x 10. We have to define two integer variables from 1 to 10 (0 being trivial). These two indexes have to vary from 1 to 10. We can begin by one loop with the index x:

```
for (int x = 1 ; x <= 10 ; x++) {

}
```

This is for the first index. The second one is totally similar:

```
for (int y = 1 ; y <= 10 ; y++) {


}
```

How can I mix those? The answer is the same as the answer to the question: what is a multiplication table? I have to keep one index constant, and multiply it by the other one going from 1 to 10. Then, I have to increment the first one and continue doing the same with the other and so on. Here is how to we do it:

```
for (int x = 1 ; x <= 10 ; x++) {

    for (int y = 1 ; y <= 10 ; y++) {

        println(x*y);
}
}
```

This code prints all results of x*y where x and y are integers from 1 to 10, one result on each line. Here are the first few steps:

- x = 1, y = 1... print the result
- x = 1, y = 2... print the result
- x = 1, y = 3... print the result

x is incremented to 2 each time the inside for loop (the one with y) ends, then x is fixed to 2 and y grows until x = 10 and y = 10 where the for loop ends.

Let's improve it a bit, only for aesthetic purposes. It is also a pretext to tweak and play with the code to make you more comfortable with it. Often, multiplication tables are drawn as follows:

1	2	3	4	5	6	7	8	9	10
2	4	6	8	10	12	14	16	18	20
3	6	9	12	15	18	21	24	27	30
4	8	12	16	20	24	28	32	36	40
5	10	15	20	25	30	35	40	45	50
6	12	18	24	30	36	42	48	54	60
7	14	21	28	35	42	49	56	63	70
8	16	24	32	40	48	56	64	72	80
9	18	27	36	45	54	63	72	81	90
10	20	30	40	50	60	70	80	90	100

Classic view of a multiplication table

We need to go to the next line each time one of the index (and only one) reaches the limit which is the value 10.

```
for (int x = 1 ; x <= 10 ; x++) {

  for (int y = 1 ; y <= 10 ; y++) {

    print(x*y);
}
println(); // add a carriage return & a new line

}
```

Check the code, each time y reaches 10, a new line is created. The for loop is a powerful structure to repeat tasks. Let's check another structure.

while loop structure

The while loop structure is a bit simpler. Here is the syntax:

```
While (expression) {
// statements
}
```

The expression is evaluated as a Boolean, true or false. While the expression is true, statements are executed, then as soon as it will be false, the loop will end. It obviously, often, requires declaration and definition outside of the while structure. Here is an example doing the same results than our first for loop printing all integers from 0 to 99:

```
int i = 0;
while (i < 100) {
println(i);
i++;
}
```

Indeed, you *have* to take care of the increment or decrement explicitly inside your statements; I'll say a few words on infinite loops a bit later. We could have condensed the code a bit more by doing that:

```
int i = 0;
while (i < 100) {
println(i++); // print the current I value, then increment i
}
```

The while loop structure tests the expression before doing even executing the first statement. Let's check a similar structure doing that differently.

do...while loop structure

The do...while loop structure is very similar to the while structure, but makes its expression evaluation at the end of the loop, which means after the statements execution. Here is the syntax:

```
do {
// statements
} while (expression);
```

Here is an example on the same model:

```
int i = 0;
do {
println(i);
i++;
} while (i < 100);
```

It means that even if the first result of the expression evaluation is false, the statements will be executed on time. This is not the case with the while structure.

Breaking the loops

We learned how to create loops driven by indexes that define precisely how these loops will live. But how can we stop a loop when an *external* event occurs? External is taken in the sense of external to the loop itself including its indexes. In that case, the loop's condition itself wouldn't include the external element.

Imagine that we have a process running 100 times in *normal* conditions. But we want to interrupt it, or modify it according to another variable that has a greater scope (declared outside of the loop, at least).

Thanks to the break statement for making that possible for us. break; is the basic syntax. When break is executed, it exits the current loop, whatever it is, based on: do, for, and while. You already saw break when we talked about the switch conditional structure. Let's illustrate that.

Imagine a LED. We want its intensity to grow from 0 to 100 percent then to go back to 0, every time. But we also want to use a nice distance sensor that resets this loop each time the distance between a user and the sensor is greater than a value.

It is based on a real installation I made for a museum where a system has to make a LED blink smoothly when the user was far and to switch off the LED when the user was near, like a living system calling for users to meet it.

I designed it very simply as follows:

```
for ( intensity = 0 ; intensity < 100 ; intensity++ ){
  ledIntensity (intensity);
  if (distance > maxDistance) { // if the user is far
    intensity = 0;    // switch off the LED
    break;        // exits the loop
  }
}
```

This whole loop was included inside the global `loop()` function in the Arduino board and the complete test about the distance was executed each time the `loop()` function occurs, waiting for users.

Infinite loops are not your friends

Be careful of infinite loops. The problem isn't really the infinite state of loops, but the fact that a system, whatever it is including Arduino, which is running an infinite loop does only that! Nothing that is after the loop can be executed because the program won't go outside the loop.

If you understand me correctly, `loop()` — the basic Arduino core function — is an infinite loop. But it is a controlled loop well designed and Arduino core based. It can (and is) interrupted when functions are called or other special events occur, letting us, users, design what we need inside of this loop. I used to call "the event's driver and listener" because it is the place where our main program runs.

There are many ways to create infinitely looped processes. You can define a variable in `setup()`, making it grow in `loop()` and test it each time `loop()` runs in order to reset it to the initial value, for instance. It takes benefits of the already existing `loop()` loop. Here is this example in C for Arduino:

```
int i;

void setup(){
  i = 0;
}

void loop(){
if (i < threshold) i +=1 ;
else i = 0;
// some statements
}
```

This i grows from 0 to threshold – 1 then goes back to 0, grows again, infinitely, taking benefits of loop().

There are also other ways to run loops infinitely in a controlled manner that we'll see a bit later in the more advanced part of the book, but you have been warned: take care of those infinite loops.

Summary

We learned a lot of abstract things in this important chapter. From type to operator's precedencies, to conditional structure, now we are going to learn new structures and syntaxes that will help us make more efficient blocks of code and, especially, more reusable ones. We can now learn about functions. Let's dive into the next C/C++ knowledge chapters and we will be able to test our Arduino after that.

4
Improve Programming with Functions, Math, and Timing

As a digital artist, I need special conditions to be able to work. We all need our own environment and ambience to be productive. Even if each one of us has his/her own way, there are many things in common.

In this chapter, I want to give you elements that will make you more comfortable to write source code that is easily readable, reusable and, as much as possible, beautiful. Like Yin and Yang, for me there has always been a Zen-like quality to the artistic and coding sides of me. Here is where I can deliver some programming pearls of wisdom to bring peace of mind to your creative side.

We are going to learn something we have already used a bit before: functions. They contribute to improve both readability and efficiency at the same time. As we do that, we'll touch on some mathematics and trigonometry often used in many projects. We'll also talk about some approaches to calculation optimization, and we'll finish this chapter with timing-related events or actions within the Arduino's firmware.

It is going to be a very interesting chapter before the real dive into pure Arduino's projects!

Introducing functions

A function is a piece of code defined by a name and that can be reused/executed from many different points in a C program. The name of a function has to be *unique* in a C program. It is also `global`, which means, as you already read for variables, it can be used everywhere in the C program containing the function declaration/definition in its scope (see the The scope concept section in *Chapter 3, C Basics – Making You Stronger*).

A function can require special elements to be passed to it; these are called **arguments**. A function can also produce and return **results**.

Structure of a function

A function is a block of code that has a header and a body. In standard C, a function's declaration and definition are made separately. The declaration of the function is specifically called the declaration of the prototype of the function and has to be done in the **header file** (see *Chapter 2, First Contact with C*).

Creating function prototypes using the Arduino IDE

The Arduino IDE makes our life easier; it creates function prototypes for us. But in special cases, if you need to declare a function prototype, you can do that in the same code file at the beginning of the code. This provides a nice way of source code centralization.

Let's take an easy example, we want to create a function that sums two integers and produces/returns the result. There are two arguments that are integer type variables. In this case, the result of the addition of these two `int` (integer) values is also an `int` value. It doesn't have to be, but for this example it is. The prototype in that case would be:

```
int mySum(int m, int n);
```

Header and name of functions

Knowing what prototype looks like is interesting because it is similar to what we call the header. The header of a function is its first statement definition. Let's move further by writing the global structure of our function `mySum`:

```
int mySum(int m, int n) // this row is the header
{
  // between curly brackets sits the body
}
```

The header has the global form:

```
returnType functionName(arguments)
```

`returnType` is a variable type. By now, I guess you understand the `void` type better. In the case where our function doesn't return anything, we have to specify it by choosing `returnType` equals to `void`.

`functionName` has to be chosen to be easy to remember and should be as self-descriptive as possible. Imagine supporting code written by someone else. Finding `myNiceAndCoolMathFunction` requires research. On the other hand, `mySum` is self-explanatory. Which code example would you rather support?

The Arduino core (and even C) follows a naming convention called camel case. The difference between two words, because we *cannot* use the blank/space character in a function name, is made by putting the first letter of words as uppercase characters. It isn't necessary, but it is recommended especially if you want to save time later. It's easier to read and makes the function self-explanatory.

`mysum` is less readable than `mySum`, isn't it? Arguments are a series of variable declarations. In our `mySum` example, we created two function arguments. But we could also have a function without arguments. Imagine a function that you need to call to produce an action that would always be the same, not depending on variables. You'd make it like this:

```
int myAction()
{
  // between curly brackets sits the body
}
```

Variables declared within a function are known only to the function containing them. This is what's known as "scope". Such variables declared within a function cannot be accessed anywhere else, but they can be "passed". Variables which can be passed are known as **arguments**.

Body and statements of functions

As you probably intuitively understood, the body is the place where everything happens; it's where all of a function's instruction steps are constructed.

Imagine the body as a real, pure, and new block of source code. You can declare and define variables, add conditions, and play with loops. Imagine the body (of instructions) as where the sculptor's clay is shaped and molded and comes out in the end with the desired effect; perhaps in one piece or many, perhaps in identical copies, and so on. It's the manipulation of what there is, but remember: Garbage in, garbage out!

You can also, as we just introduced, return a variable's value. Let's create the body of our mySum example:

```
int mySum(int m, int n) // this row is the header
{
  int result;          // this is the variable to store the result
  result = m + n;      // it makes the sum and store it in result
  return result; // it returns the value of result variable
}
```

int result; declares the variable, and names it result. Its scope is the same as the scope of arguments. result = m + n; contains two operators, and you already know that + has a higher precedence than = which is pretty good, as the mathematical operation is made first and then the result is stored in the result variable. This is where the magic happens; take two operators, and make one out of them. Remember that, in a combination of multiple mathematical operations, do not forget the order of precedence; it's critical so that we don't get unexpected results.

At last, return result; is the statement that makes the function call resulting into a value. Let's check an actual example of the Arduino code to understand this better:

```
void setup() {
Serial.begin(9600); Let's check an actual example of Arduino code to
understand this better.
}

Void loop() {
// let's sum all integers from 0 to 99, 2 by 2 and display
int currentResult;
for (int i = 0 ; i < 100 ; i++)
{
  currentResult = mySum(i,i+1);   // sum and store
  Serial.println(currentResult);   // display to serial monitor
}
delay(2000); make a 2 second pause
}
```

```
int mySum(int m, int n) // this row is the header
{
   int result;            // this is the variable to store the result
   result = m + n;        // it makes the sum and store it in result
   return result; // it returns the value of result variable
}
```

As you have just seen, the `mySum` function has been defined and called in the example. The most important statement is `currentResult = mySum(i,i+1);`. Of course, the `i` and `i+1` trick is interesting, but the most important thing to recognize here is the usage of the variable `currentResult` that was declared at the beginning of the `loop()` function.

In programming, it's important to recognize that everything at the right (contents) goes into the left (the new container). According to the precedence rules, a function call has a precedence of 2 against 16 for the = assignment operator. It means the call is made first and the function returns the result of the + operation, as we designed it. From this point of view, you just learned something very important: *The call statement of a function returning a result is a value.*

You can check *Appendix B, Operator Precedence in C and C++* for the entire precedencies list. As with all values within a variable, we can store it into another, here inside the integer variable `result`.

Benefits of using functions

Programming is about writing pieces of code for general and specific purposes. Using functions is one of the best ways of segmenting your code.

Easier coding and debugging

Functions can really help us to be better organized. While designing the program, we often use pseudo-codes and this is also the step when we notice that there are a lot of common statements. These common statements may often be put inside functions.

The function/call pattern is also easier to debug. We have only one part of code where the function sits. If there is a problem, we can debug the function itself just once, and all the calls will then be fixed instead of modifying the whole part of the code.

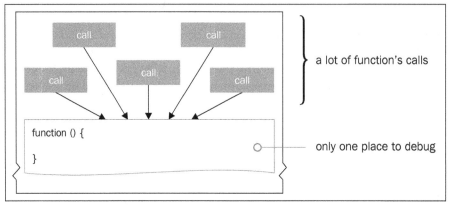

Functions make your code easier to debug

Better modularity helps reusability

Some part of your code will be high level and general. For instance, at some point, you may need a series of statements that can cut an array into pieces, then regroup all values following a basic rule. This series could be the body of a function. In another way, coding a function that converts Fahrenheit units into Celsius could interest you. These two examples are general-purpose functions.

In contrast, you can also have a specific function whose sole purpose is to convert U.S. Dollars to French Francs. You may not call it very often, but if occasionally necessary, it is always ready to handle that task.

In both cases, the function can be used and of course, re-used. The idea behind this is to save time. It also means that you can grab some already existing functions and re-use them. Of course, it has to be done following some principles, such as:

- Code licensing
- Respect the API of the function that can be a part of a library
- Good match for your purpose

Code licensing issue is an important point. We are used to grabbing, testing, and copy/pasting things, but the code you find isn't always in the public domain. You have to take care of the license file that is often included in a code release archive, and in the first line of the code too, where comments can help you understand the conditions to respect the re-use of it.

Application Programming Interface (API) means you have to conform yourself to some documentation before using the material related to that API. I understand that purists would consider this a small abuse, but it is a pretty pragmatic definition.

Basically, an API defines specifications for routines, data structures, and other code entities that can be re-used inside other programs. An API specification can be documentation of a library. In that case, it would precisely define what you can and cannot do.

The good-match principle can seem obvious, but sometimes out of convenience we find an existing library and choose to use it rather than coding our own solution. Unfortunately, sometimes in the end we only add more complication than originally intended. Doing it ourselves may fulfil the simple need, and will certainly avoid the complexities and idiosyncrasies of a more comprehensive solution. There's also the avoidance of a potential performance hit; you don't buy a limo when all you really need is to walk to the supermarket just down the street.

Better readability

It is a consequence of the other benefits, but I want to make you understand that this is more vital than commenting your code. Better readability means saving time to focus on something else. It also means easier code upgrade and improvement steps.

C standard mathematical functions and Arduino

As we have already seen, almost all standard C and C++ entities supported by the compiler **avr-g++** should work with Arduino. This is also true for C mathematical functions.

This group of functions is a part of the (famous) C standard library. A lot of functions of this group are inherited in C++. There are some differences between C and C++ in the use of complex numbers. C++ doesn't provide complex numbers handling from that library but from its own C++ standard library by using the class template `std::complex`.

Almost all these functions are designed to work with and manipulate floating-point numbers. In standard C, this library is known as `math.h` (a filename), which we mention in the header of a C program, so that we can use its functions.

Trigonometric C functions in the Arduino core

We often need to make some trigonometric calculations, from determining distances an object has moved, to angular speed, and many other real-world properties. Sometimes, you'll need to do that inside Arduino itself because you'll use it as an autonomous smart unit without any computers in the neighborhood.

The Arduino core provides the classic trigonometric functions that can be summarized by writing their prototypes. A major part of these return results in radians. Let's begin by reviewing our trigonometry just a bit!

Some prerequisites

I promise, I'll be quick and light. But the following lines of text will save you time looking for your old and torn school book. When I learn knowledge from specific fields, I personally like to have all I need close at hand.

Difference between radians and degrees

Radian is the unit used by many trigonometric functions. Then, we have to be clear about radians and degrees, and especially how to convert one into the other. Here is the official radian definition: **Alpha** is a ratio between two distances and is in radian units.

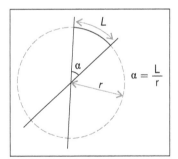

Radian definition

Degree is the 1/360 of a full rotation (complete circle). Considering these two definitions and the fact that a complete rotation equals 2π, we can convert one into the other:

[
angleradian = angledegree x π/180
angledegree = angleradian x 180/π
]

Cosine, sine, and tangent

Let's see the trigonometric triangle example:

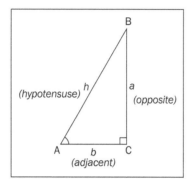

Considering the angle A in radians, we can define cosine, sine, and tangent as follows:

- cos(A) = b/h
- sin(A) = a/h
- tan(A) = sin(A)/cos(A) = a/b

Cosine and sine evolve from -1 to 1 for value of angles in radians, while tangent has some special points where it isn't defined and then evolves cyclically from -∞ to +∞. We can represent them on the same graph as follows:

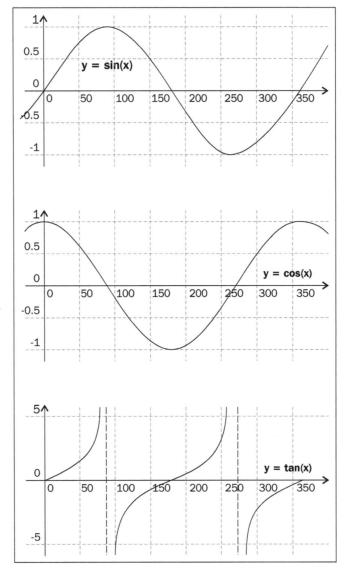

Graphical cosine, sine, and tangent representation

Yes, of course, those functions oscillate, infinitely reproducing the same evolutions. It is good to keep in mind that they can be used for pure calculations but also to avoid overly linear value evolution in the time by replacing linearity by smoother oscillations. We'll see that a bit later.

We know how to calculate a cosine/sine/tangent when we have an angle, but how to calculate an angle when we already have the cosine/sine/tangent?

Arccosine, arcsine, and arctangent

Arccosine, arcsine, and arctangent are called inverse trigonometric functions. These functions are used to calculate an angle when you already have the distance ratios that I mentioned before.

They are called inverse because this is the inverse/reciprocal process of the previously seen trigonometric function. Basically, these functions provide you an angle, but considering the periodicity, they provide a lot of angles. If k is an integer, we can write:

- $\sin (A) = x$ ó $A = \arcsin(x) + 2k\pi$ or $y = \pi - \arcsin(x) + 2k\pi$
- $\cos (A) = x$ ó $A = \arccos(x) + 2k\pi$ or $y = 2\pi - \arccos (x) + 2k\pi$
- $\tan (A) = x$ ó $A = \arctan(x) + k\pi$

These are the right mathematical relationships. Practically, in usual cases, we can drop the full rotation cases and forget about the $2k\pi$ of the cosine and sine cases and $k\pi$ of the tangent case.

Trigonometry functions

`Math.h` contains the trigonometry function's prototype, so does the Arduino core:

- `double cos (double x);` returns the cosine of x radians
- `double sin (double x);` returns the sine of x radians
- `double tan (double x);` returns the tangent of x radians
- `double acos (double x);` returns A, the angle corresponding to $\cos (A) = x$
- `double asin (double x);` returns A, the angle corresponding to $\sin (A) = x$
- `double atan (double x);` returns A, the angle corresponding to $\tan (A) = x$
- `double atan2 (double y, double x);` returns $\arctan (y/x)$

Exponential functions and some others

Making calculations, even basic ones, involves other types of mathematical functions, namely power, absolute value, and so on. The Arduino core then implements those. Some mathematical functions are given as follows:

- `double pow (double x, double y);` returns x to power y
- `double exp (double x);` returns the exponential value of x
- `double log (double x);` returns the natural logarithm of x with x > 0
- `double log10 (double x);` returns the logarithm of x to base 10 with x > 0
- `double square (double x);` returns the square of x
- `double sqrt (double x);` returns the square root of x with x >= 0
- `double fabs (double x);` returns the absolute value of x

Of course, mathematical rules, especially considering range of values, have to be respected. This is why I added some conditions of x to the list.

All these functions are very useful, even for solving small problems. One day, I was teaching someone at a workshop and had to explain about measuring temperature with a sensor. This student was quite motivated but didn't know about these functions because she only played with inputs and outputs without converting anything (because she basically didn't need that). We then learned these functions, and she ended by even optimizing her firmware, which made me so proud of her!

Now, let's approach some methods of optimization.

Approaching calculation optimization

This section is an approach. It means it doesn't contain all the advanced tips and tricks for programming optimizations, but contains the optimizations on pure calculation.

Generally, we design an idea, code a program, and then optimize it. It works fine for huge programs. For smaller ones, we can optimize while coding.

 Normally, our firmware is small and so I'd suggest that you consider this as a new rule: Write each statement keeping optimization in mind.

I could add something else right now: Don't kill the readability of your code with too many cryptic optimization solutions; I thought of *pointers* while writing that. I'll add a few lines about them in order to make you familiar with, at least, the concept.

The power of the bit shift operation

If I consider an array to store things, I almost always choose the size as a power of two. Why? Because the compiler, instead of performing the array indexing by using a CPU-intensive multiply operation, can use the more efficient bit shift operation.

What are bit operations?

Some of you must have already understood the way I work; I'm using a lot of pretexts to teach you new things. Bitwise operators are specific operators for bits. Some cases require this kind of calculation. I can quote two cases that we'll learn about in the next part of this book:

- Using shift registers for multiplexing
- Performing arithmetic operations, for powers of 2, involving the multiply and divide operator

There are four operators and two bit shift operators. Before we dive into it, let's learn a bit more about the binary numeral system.

Binary numeral system

We are used to counting using the decimal system, also called decimal numeral system or base-10 number system. In this system, we can count as:

0, 1, 2, 3, 4, 5, 6, 7, 8, 9, 10, 11, 12...

Binary numeral system is the system used under the hood in computers and digital electronic devices. It is also named base-2 system. In this system we count as follows:

0, 1, 10, 11, 100, 101, 110, 111...

Easily converting a binary number to a decimal number

A nice trick to convert from binary to decimal, start by counting the position of 0 and 1, starting from the index 0.

Let's take 110101. It can be represented as follows:

Positions	0	1	2	3	4	5
	1	0	1	0	1	1

Then, I can write this sum of multiplications and it equals the decimal version of my 110101 number:

1 x 20 + 0 x 21 + 1 x 22 + 0 x 23 + 1 x 24 + 1 x 25 = 1 + 4 + 16 + 32 = 53

Each bit *decides* if we have to consider the power of 2, considering its position.

AND, OR, XOR, and NOT operators

Let's have a look at these four operators.

AND

The bitwise AND operator is written with a single ampersand: &. This operator operates on each bit position independently according to the following rules:

- 0 & 0 == 0
- 0 & 1 == 0
- 1 & 0 == 0
- 1 & 1 == 1

Let's take a real example with integers, which are a 16-bit value:

```
int a = 35;     // in binary: 00000000 00100011
int b = 49;     // in binary: 00000000 00110001
int c = a & b;  // in binary: 00000000 00100001 and in decimal 33
```

To find the result easily, we have to compare each bit one by one for each position while following the preceding rules.

OR

The bitwise OR operator is written with a single vertical bar: |. It can be done by pressing *Alt + Shift + l* (letter L) on OSX and *Shift + * on other PC keyboards. This operator operates on each bit position independently according to the following rules:

- 0 | 0 == 0
- 0 | 1 == 1
- 1 | 0 == 1
- 1 | 1 == 1

XOR

The bitwise XOR operator is written with a single caret symbol: ^. This operator operates on each bit position independently according to the following rules:

- 0 ^ 0 == 0

- 0 ^ 1 == 1

- 1 ^ 0 == 1

- 1 ^ 1 == 0

It is the exclusive version of OR, thus the name XOR.

NOT

The bitwise XOR operator is written with a tilde symbol: ~. It is a unary operator, which means, if you remember this term correctly, it can apply to one number only. I call it the *bit changer* in my workshops. It changes each bit to its opposite:

- ~0 == 1

- ~1 == 0

Let's take a real example with integers, which are 16-bit values as you know:

```
int a = 35;    // in binary: 00000000 00100011
int b = ~a ;   // in binary: 11111111 11011100 and in decimal -36
```

As you already know, the int type in C is a signed type (*Chapter 3, C Basics – Making You Stronger*) that is able to encode numbers from -32,768 to 32,767 – negative numbers too.

Bit shift operations

Coming from C++, the left shift and the right shift operators are respectively symbolized by << and >>. It is easy to remember, the double << goes to the left, and the other one >> to the right. Basically, it works like this:

```
int a = 36;              // in binary 00000000 00100100
int b = a << 2;  // in binary 00000000 10010000, decimal 144
int c = a >> 1;  // in binary 00000000 00010010, decimal 18
```

It is quite easy to see how it works. You shift all bits from a particular number of positions to the left or to the right. Some of you would have noticed that this is the same as multiplying or dividing by 2. Doing << 1 means multiply by 2, >> 1 means divide by 2. << 3 means multiply by 8 (23), >> 5 means divide by 32 (25), and so on.

It is all about performance

Bitwise operations are primitive actions directly supported by the processor. Especially with embedded systems, which are still not as powerful as normal computers, using bitwise operations can dramatically improve performance. I can write two new rules:

- Using power of 2 as the array size drives the use of bit shift operators internally/implicitly while the CPU performs index calculations. As we just learned, multiplying/dividing by 2 can be done very efficiently and quickly with bit shift.

- All your multiplications and divisions by a power of 2 should be replaced by bit shifting.

This is the nicest compromise between cryptic code and an efficient code. I used to do that quite often. Of course, we'll learn real cases using it. We are still in the most theoretical part of this book, but everything here will become clear quite soon.

The switch case labels optimization techniques

The `switch...case` conditional structure can also be optimized while you are writing it.

Optimizing the range of cases

The first rule is to place all cases of the considered switch in the narrowest range possible.

In such a case, the compiler produces what we call a *jump table of case labels*, instead of generating a huge `if-else-if` cascade. The jump table based `switch... case` statement 's performance is independent of the number of case entries in the `switch` statement.

[So, place all cases of the switch in the narrowest range possible.]

Optimizing cases according to their frequency

The second rule is to place all cases sorted from the most frequently occurring to the least frequently occurring when you know the frequency.

As mentioned before, in cases where your `switch` statement contains cases placed far apart, because you cannot handle that in another way, the compiler replaces the `switch` statement and generates `if-else-if` cascades. It means it will always be better to reduce the potential number of comparisons; this also means that if the cases that are most probable are placed at the beginning, you maximize your chances to do that.

 So, place all cases sorted from the most frequently occurring to the least frequently occurring.

Smaller the scope, the better the board

As I already mentioned when we talked about a variables' scope, always use the smallest scope possible for any variables. Let's check this example with a function named `myFunction`:

```
int myFunction( int valueToTest )
{
  if (valueToTest == 1)
  {
    int temporaryVariable;
    // some calculations with temporaryVariable
    return temporaryVariable;
  }
  else {
  return -1;
  }
}
```

`temporaryVariable` is only required in one case, when `valueToTest` equals 1. If I declare `temporaryVariable` outside of the `if` statement, whatever the value of `valueToTest`, `temporaryVariable` will be created.

In the example I cite, we save memory and processing; in all cases where `valueToTest` is not equal to 1, the variable `temporaryVariable` is not even created.

 Use the smallest scope possible for all your variables.

The Tao of returns

Functions are usually designed with a particular idea in mind, they are modules of code able to perform specific operations through the statements that they include and are also able to return a result. This concept provides a nice way to forget about all those specific operations performed inside the function when we are outside of the function. We know the function has been designed to provide you a result when you give arguments to it.

Again, this is a nice way to focus on the core of your program.

Direct returns concept

As you may have already understood, declaring a variable creates a place in memory. That place cannot be used by something else, of course. The process that creates the variable consumes processor time. Let's take the same previous example detailed a bit more:

```
int myFunction( int valueToTest )
{
  if (valueToTest == 1)
  {
     int temporaryVariable;
     temporaryVariable += globalVariable;
     temporaryVariable *= 7;
     return temporaryVariable;
  }
  else {
  return -1;
  }
}
```

What could I improve to try to avoid the use of `temporaryVariable`? I could make a *direct return* as follows:

```
int myFunction( int valueToTest )
{
  if (valueToTest == 1)
  {
     return ( (globalVariable + 1)*7 );
  }
  else {
  return -1;
  }
}
```

In the longer version:

- We were inside the `valueToTest == 1` case thus `valueToTest` equals 1
- I directly put the calculation in the `return` statement

In that case, there is no more temporary variable creation. There are some cases where it can be more readable to write a lot of temporary variables. But now, you are aware that it is worth finding compromises between readability and efficiency.

 Use a direct return instead of a lot of temporary variables.

Use void if you don't need return

I often read code including functions with a return type that didn't return anything. The compiler may warn you about that. But in case it didn't, you have to take care of it. A call to a function that provides a return type will always pass the return value even if nothing inside the function's body is really returned. This has a CPU cost.

 Use void as a return type for your functions if they don't return anything.

Secrets of lookup tables

Lookup tables are one of the most powerful tricks in the programming universe. They are arrays containing precalculated values and thus replace heavy runtime calculations by a simpler array index operation. For instance, imagine you want to track positions of something by reading distances coming from a bunch of distance sensors. You'll have *trigonometric* and probably *power* calculations to perform. Because they can be time consuming for your processor, it would be smarter and cheaper to use array content reading instead of those calculations. This is the usual illustration for the use of lookup tables.

These lookup tables can be precalculated and stored in a static program's storage memory, or calculated at the program's initialization phase (in that case, we call them *prefetched lookup tables*).

Some functions are particularly expensive, considering the CPU work. Trigonometric functions are one such function that can have bad consequences as the storage space and memory are limited in embedded systems. They are typically prefetched in code. Let's check how we can do that.

Table initialization

We have to precalculate the cosine **Look Up Table** (**LUT**). We need to create a small precision system. While calling cos(x) we can have all values of x that we want. But if we want to prefetch values inside an array, which has by design a finite size, we have to calculate a finite number of values. Then, we cannot have our cos(x) result for all float values but only for those calculated.

I consider precision as an angle of 0.5 degrees. It means, for instance, that the result of cosine of 45 degrees will be equal to the cosine of 45 degrees 4 minutes in our system. Fair enough.

Let's consider the Arduino code. You can find this code in the Chapter04/CosLUT/ folder:

```
float cosLUT[(int) (360.0 * 1 / 0.5)] ;
const float DEG2RAD = 180 / PI ;


const float cosinePrecision = 0.5;
const int cosinePeriod = (int) (360.0 * 1 / cosinePrecision);


void setup()
{
  initCosineLUT();
}

void loop()
{
  // nothing for now!
}


void initCosineLUT(){
  for (int i = 0 ; i < cosinePeriod ; i++)
  {
    cosLUT[i] = (float) cos(i * DEG2RAD * cosinePrecision);
  }
}
```

cosLUT is declared as an array of the type float with a special size. 360 * 1/ (precision in degrees) is just the number of elements we need in our array considering the precision. Here, precision is 0.5 degrees and of course, the declaration could be simplified as follows:

```
float cosLUT[720];
```

We also declared and defined a `DEG2RAD` constant that is useful to convert degrees to radians. We declared `cosinePrecision` and `cosinePeriod` in order to perform those calculations once.

Then, we defined an `initCosineLUT()` function that performs the precalculation inside the `setup()` function. Inside its body, we can see a loop over index `i`, from `i=0` to the size of the array minus one. This loop precalculates values of cosine(x) for all values of x from 0 to 2π. I explicitly wrote the x as `i * DEG2RAD * precision` in order to keep the precision visible.

At the board initialization, it calculates all the lookup table values once and provides these for further calculation by a simple array index operation.

Replacing pure calculation with array index operations

Now, let's retrieve our cosine values. We can easily retrieve our values by accessing our LUT through another function, shown as follows:

```
float myFastCosine(float angle){

    return cosLUT[(int) (angle * 1 / cosinePrecision) % cosinePeriod];
}
```

`angle * 1 / cosinePrecision` gives us the angle considering the given precision of our LUT. We apply a modulo operation considering the `cosinePeriod` value to wrap values of higher angles to the limit of our LUT, and we have our index. We directly return the array value corresponding to our index.

We could also use this technique for root square prefetching. This is the way I used it in another language when I coded my first iOS application named **digital collisions** (`http://julienbayle.net/blog/2012/04/07/digital-collisions-1-1-new-features`). If you didn't test it, this is an application about generative music and visuals based on physical collision algorithms. I needed a lot of distance and rotation calculations. Trust me, this technique turned the first sluggish prototype into a fast application.

Taylor series expansion trick

There is another nice way to save CPU work that requires some math. I mean, a bit more advanced math. Following words are very simplified. But yes, we need to focus on the C side of things, and not totally on mathematics.

Taylor series expansions are a way to approximate almost every mathematical expression at a particular point (and around it) by using polynomial expressions.

 A polynomial expression is similar to the following expression:
$P(x) = a + bx + cx2 + dx3$

$P(x)$ is a polynomial function with a degree 3. a, b, c, and d are float numbers.

The idea behind a Taylor series is that we can approximate an expression by using the first term of the theoretical infinite sums that represent this expression. Let's take some examples.

For instance, considering x evolving from -π and π; we can write the function sine as follows:

$sin(x) \approx x - x3/6 + x5/120 - x7/5040$

The sign ≈ means "approximately equals". Inside a reasonable range, we can replace $sin(x)$ by $x - x3/6 + x5/120 - x7/5040$. There is no magic, just mathematical theorems. We can also write x evolving from -2 to 3 as follows:

$ex \approx 1 + x + x2/2 + x3/6 + x4/24$

I could add some other examples here, but you'll be able to find this in the *Appendix D, Some Useful Taylor Series for Calculation Optimization*. These techniques are some powerful tricks to save CPU time.

The Arduino core even provides pointers

Pointers are more complicated techniques for beginners in C programming but I want you to understand the concept. They are not data, but they point to the starting point of a piece of data. There are at least two ways to pass data to a function or something else:

- Copy it and pass it
- Pass a pointer to it

In the first case, if the amount of data is too large our memory stack would explode because the whole data would be copied in the stack. We wouldn't have a choice other than a pointer pass.

In this case, we have the reference of the place where the data is stored in memory. We can operate exactly as we want but only by using pointers. Pointers are a smart way to deal with any type of data, especially arrays.

Time measure

Time is always something interesting to measure and to deal with, especially in embedded software that is, obviously, our main purpose here. The Arduino core includes several time functions that I'm going to talk about right now.

There is also a nice library that is smartly named **SimpleTimer Library** and designed as a GNU LGPL 2.1 + library by *Marcello Romani*. This is a good library based on the `millis()` core function which means the maximum resolution is 1 ms. This will be more than enough for 99 percent of your future projects. *Marcello* even made a special version of the library for this book, based on `micros()`.

The Arduino core library now also includes a native function that is able to have a resolution of 8 microseconds, which means you can measure time delta of 1/8,000,000 of a second; quite precise, isn't it?

I'll also describe a higher resolution library **FlexiTimer2** in the last chapter of the book. It will provide a high-resolution, customizable timer.

Does the Arduino board own a watch?

The Arduino board chip provides its *uptime*. The *uptime* is the time since the board has started. It means you cannot natively store absolute time and date without keeping the board up and powered. Moreover, it will require you to set up the absolute time once and then keep the Arduino board powered. It is possible to keep the board autonomously powered. I also talk about that later in this book.

The millis() function

The core function `millis()` returns the number of milliseconds since the board has been started the last time. For your information, 1 millisecond equals 1/1000 of a second.

The Arduino core documentation also provides that this number will go back to zero after approximately 50 days (this is called the timer overflow). You can smile now, but imagine your latest installation artistically illustrating the concept of time in the MoMA in NYC which, after 50 days, would get totally messed up. You would be interested to know this information, wouldn't you? The return format of `millis()` is *unsigned long*.

Here is an example you'll have to upload to your board in the next few minutes. You can also find this code in the `Chapter04/ measuringUptime/` folder:

```
/*
   measuringTime is a small program measuring the uptime and printing
it
   to the serial monitor each 250ms in order not to be too verbose.

   Written by Julien Bayle, this example code is under Creative Commons
CC-BY-SA

   This code is related to the book "C programming for Arduino" written
by Julien Bayle
   and published by Packt Publishing.

   http://cprogrammingforarduino.com
 */

unsigned long measuredTime;        // store the uptime

void setup(){
  Serial.begin(9600);
}

void loop(){
  Serial.print("Time: ");
  measuredTime = millis();

  Serial.println(measuredTime);   // prints the current uptime

  delay(250);              // pausing the program 250ms
}
```

Can you optimize this (only for pedagogical reasons as this is a very small program)? Yes, indeed, we can avoid the use of the `measuredTime` variable. It would look more like this:

```
/*
   measuringTime is a small program measuring the uptime and printing
it
   to the serial monitor each 250ms in order not to be too verbose.

   Written by Julien Bayle, this example code is under Creative Commons
CC-BY-SA
```

```
   This code is related to the book "C programming for Arduino" written
by Julien Bayle
   and published by Packt Publishing.

   http://cprogrammingforarduino.com
 */

void setup(){
  Serial.begin(9600);
}

void loop(){
  Serial.print("Time: ");
  Serial.println(millis());  // prints the current uptime
  delay(250);                // pausing the program 250ms
}
```

It is also beautiful in its simplicity, isn't it? I'm sure you'll agree. So upload this code on your board, start the Serial Monitor, and look at it.

The micros() function

If you needed more precision, you could use the `micros()` function. It provides uptime with a precision of 8 microseconds as written before but with an overflow of approximately 70 minutes (significantly less than 50 days, right?). We gain precision but loose overflow time range. You can also find the following code in the `Chapter04/measuringUptimeMicros/` folder:

```
/*
   measuringTimeMicros is a small program measuring the uptime in ms
and
   µs and printing it to the serial monitor each 250ms in order not to
be too verbose.

   Written by Julien Bayle, this example code is under Creative Commons
CC-BY-SA

   This code is related to the book «C programming for Arduino» written
by Julien Bayle
   and published by Packt Publishing.

   http://cprogrammingforarduino.com
 */

void setup(){
```

```
    Serial.begin(9600);
}

void loop(){
  Serial.print(«Time in ms: «);
  Serial.println(millis());   // prints the current uptime in ms
  Serial.print(«Time in µs: «);
  Serial.println(micros());   // prints the current uptime in µs

  delay(250);                 // pausing the program 250ms
}
```

Upload it and check the Serial Monitor.

Delay concept and the program flow

Like Le Bourgeois Gentilhomme who spoke prose without even realizing it, you've already used the `delay()` core function and haven't realized it. Delaying an Arduino program can be done using the `delay()` and `delayMicroseconds()` functions directly in the `loop()` function.

Both functions drive the program to make a pause. The only difference is that you have to provide a time in millisecond to `delay()` and a time in microseconds to `delayMicroseconds()`.

What does the program do during the delay?

Nothing. It waits. This sub-subsection isn't a joke. I want you to focus on this particular point because later it will be quite important.

 When you call `delay` or `delayMicroseconds` in a program, it stops its execution for a certain amount of time.

Here is a small diagram illustrating what happens when we power on our Arduino:

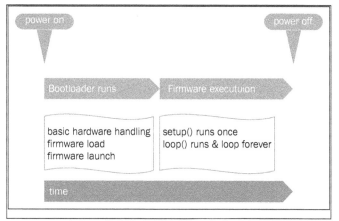

One lifecycle of a Arduino's firmware

Now here is a diagram of the firmware execution itself, which is the part that we will work with, in the next rows:

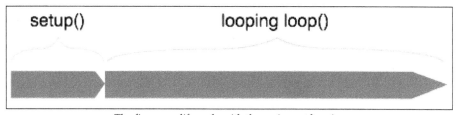

The firmware life cycle with the main part looping

Accepting the fact that when setup() stops, the loop() function begins to loop, everything in loop() is continuous. Now look at the same things when delays happen:

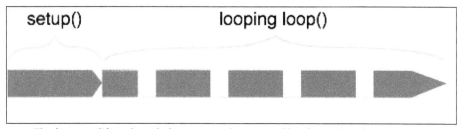

The firmware life cycle with the main part looping and breaking when delay() is called

The whole program breaks when delay() is called. The length of the break depends on the parameter passed to delay().

We can notice that everything is done sequentially and in time. If a statement execution takes a lot of time, Arduino's chip executes it, and then continues with the next task.

In that very usual and common case, if one particular task (statements, function calls, or whatever) takes a lot of time, the whole program could be hung and produce a hiccup; consider the user experience.

Imagine that concrete case in which you have to read sensors, flip-flop some switches, and write information to a display *at the same time*. If you do that sequentially and you have a lot of sensors, which is quite usual, you can have some lag and slowdown in the display of information because that task is executed after the other one in loop().

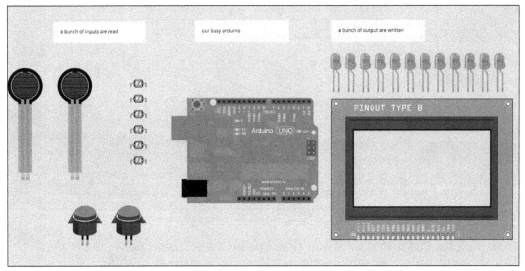

An Arduino board busy with many inputs and outputs

I usually teach my students at least two concepts in dealing with that only-one-task property that can feel like a limitation:

- Thread
- Interrupt handler (and subsequent interrupt service routine concept)

I obviously teach another one: *The polling*. **The polling is a special interrupt case from where we will begin.**

The polling concept – a special interrupt case

You know the poll term. I can summarize it as "ask, wait for an answer, and keep it somewhere".

If I wanted to create a code that reads inputs, and performs something when a particular condition would be verified with the value of these inputs, I would write this pseudo-code:

```
setup()
- initialize things

loop()
- ask an input value and wait for it until it is available
- test this input according to something else
- if the test is true perform something else, loop to the beginning
```

What could be annoying here? I cyclically poll new information and have to wait for it.

During this step, nothing more is done, but imagine that the input value remains the same for a long time. I'd request this value cyclically in the loop, constraining the other tasks to wait for nothing.

It sounds like a waste of time. Normally, polling is completely sufficient. It has to be written here instead of what other raw programmers could say to you.

We are creators, we need to make things communicate and work, and we can and like to test, don't we? Then, you just learned something important here.

 Don't design complex program workarounds before having tested basic ones.

One day, I asked some people to design basic code. Of course, as usual, they were connected to the Internet and I just agreed because we are almost all working like that today, right? Some people finished before others.

Why? A lot of the people who finished later tried to build a nice multithreaded workaround using a messaging system and an external library. The intention was good, but in the time we had, they didn't finish and only had a nice Arduino board, some wired components, and a code that wasn't working on the table.

Do you want to know what the others had on their desktop? A polling-based routine that was driving their circuits perfectly! Time wasted by this polling-based firmware was just totally unimportant considering the circuit.

[Think about hardcore optimizations, but first test your basic code.]

The interrupt handler concept

Polling is nice but a bit time consuming, as we just figured out. The best way would be to be able to control when the processor would have to deal with inputs or outputs in a smarter way.

Imagine our previously drawn example with many inputs and outputs. Maybe, this is a system that has to react according to a user action. Usually, we can consider the user inputs as much slower than the system's ability to answer.

This means we could create a system that would interrupt the display as soon as a particular event would occur, such as a user input. This concept is called an *event-based interrupt system*.

The *interrupt* is a signal. When a particular event occurs, an interrupt message is sent to the processor. Sometimes it is sent externally to the processor (hardware interrupt) and sometimes internally (software interrupt).

This is how the disk controller or any external peripheral informs the processor of the main unit that it has to provide this or that at the right moment.

The interrupt handler is a routine that handles the interrupt by doing something. For instance, on the move of the mouse, the computer operating system (commonly called the OS) has to redraw the cursor in another place. It would be crazy to let the processor itself test each millisecond whether the mouse has moved, because the CPU would be running at 100 percent utilization. It seems smarter to have a part of the hardware for that purpose. When the mouse movement occurs, it sends an interrupt to the processor, and this later redraws the mouse.

In the case of our installation with a huge number of inputs and outputs, we can consider handling the user inputs with an interrupt. We would have to implement what is called an **Interrupt Service Routine (ISR)**, which is a routine called only when a physical world event occurs, that is, when a sensor value changes or something like that.

Arduino now provides a nice way to attach an interrupt to a function and it is now easy to design an ISR (even if we'll learn to do that a bit later). For instance, we can now react to the change of the value of an analog thermal sensor using ISR. In this case, we won't permanently poll the analog input, but we'll let our low-level Arduino part do that. Only when a value changes (rises or falls) depending on how we have attached the interrupt, would this act as a trigger and a special function would execute (for instance, the LCD display gets updated with the new value).

Polling, ISR, and now, we'll evoke threads. Hang on!

What is a thread?

A thread is a running program flow in which the processor executes a series of tasks, generally looping, but not necessarily.

With only one processor, it is usually done by *time-division multiplexing*, which means the processor switches between the different threads according to time, that is, context switching.

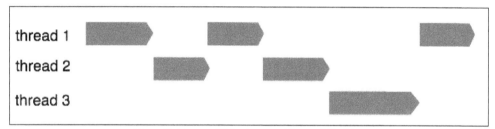

Time-division multiplexing provides multitasking

More advanced processors provide the *multithread* feature. These behave as if they would be more than just one, each part dealing with a task at the same time.

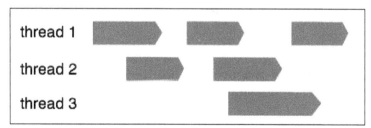

Real multithreading provides tasks happening at the same time

Without going deeper into computer processors, as we aren't dealing with them right now, I can say threads are nice techniques to use in programming to make tasks run simultaneously.

Unfortunately, the Arduino core doesn't provide multithreading, nor does any other microcontroller. Because Arduino is an open source hardware project, some hackers have designed a variant of the Arduino board and created some Freeduino variant providing *concurrency*, an open source programming language, and an environment designed especially with multithreading in mind. This is out of topic here, but at least, you now have some leads if you are interested.

Let's move to the second solution to go beyond the one-task-at-a-time constraint, if we need it.

A real-life polling library example

As introduced in the first line of this section, Marcello's library is a very nice one. It provides a polling-based way to launch timed actions.

Those actions are generally function calls. Functions that behave like that are sometimes known as callback functions. These functions are generally called as an argument to another piece of code.

Imagine that I want to make our precious LED on the Arduino board blink every 120 milliseconds. I could use a delay but it would totally stop the program. Not smart enough.

I could hack a hardware timer on the board, but that would be overkill. A more practical solution that I would use is a callback function with Marcello's `SimpleTimer` library. Polling provides a simple and inexpensive way (computationally speaking) to deal with applications that are not timer dependent while avoiding the use of interrupts that raise more complex problems like hardware timer overconsumption (hijacking), which leads to other complicated factors.

However, if you want to call a function every 5 milliseconds and that function needs 9 milliseconds to complete, it will be called every 9 milliseconds. In our case here, with 120 milliseconds required to produce a nice and eye-friendly, visible blink, we are very safe.

For your information, you don't need to wire anything more than the USB cable between the board and your computer. The board-soldered LED on Arduino is wired to digital pin 13. Let's use it.

But first, let's download the `SimpleTimer` library for your first use of an external library.

Installing an external library

Download it from `http://playground.arduino.cc/Code/SimpleTimer`, and extract it somewhere on your computer. You will typically see a folder with at least two files inside:

- A header file (`.h` extension)
- A source code file (`.cpp` extension)

Now, you can see for yourself what they are. Within these files, you have the source code. Open your sketchbook folder (see *Chapter 1*, *Let's Plug Things*), and move the library folder into the `libraries` folder if it exists, else create this special folder:

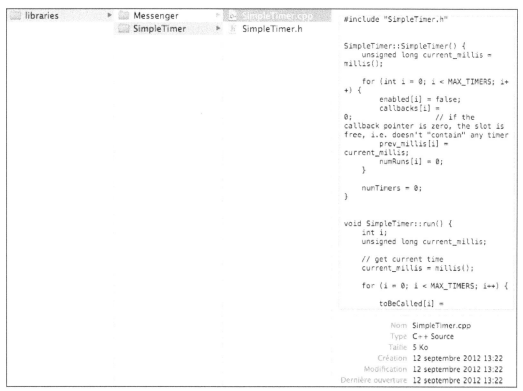

The header and the source code of SimpleTimer by Marcello Romani

The next time you'll start your Arduino IDE, if you go to **Sketch | Import Library**, you'll see a new library at the bottom.

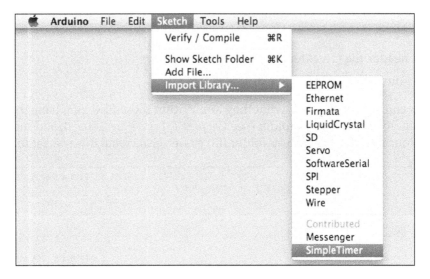

In order to include a library, you can click on it in this menu and it will write `#include <libname.h>` in your code. You can also type this by yourself.

Let's test the code

Upload this next code and reboot Arduino; I'm going to explain how it works. You can also find this code in the `Chapter04/simpleTimerBlinker/` folder:

```
#include <SimpleTimer.h>         // include the Marcello's library

SimpleTimer timer ;              // the timer object construction
boolean currentLEDState ;
int ledPin = 13 ;


void setup() {
currentLEDState = false ;
pinMode(ledPin, OUTPUT) ;
timer.setInterval(120, blink) ;

}

void loop() {
timer.run() ;
}
```

```
// a function to be executed periodically
void blink() {
  if (!currentLEDState) digitalWrite(ledPin, HIGH);
else digitalWrite(ledPin, LOW);
currentLEDState = !currentLEDState ; // invert the boolean
}
```

This library is easy to use in our case. You have to include it first, of course. Then you have to declare an instance of SimpleTimer, which is an object construct, by declaring it.

Then I'm using a currentLEDState Boolean value to store the current state of the LED explicitly. At last, I declare/define ledPin with the number of the pin I need (in this case, 13) to make the LED blink. setup() is basically done with some initialization. The most important one here is the timer.setInterval() function.

Maybe, this is your first method call. The object timer has and embeds some methods that we can use. One of them is setInterval, which takes two variables:

- A time interval
- A callback function

We are passing a function name here (a piece of code) to another piece of code. This is the structure of a typical callback system.

loop() is then designed by calling the run() method of the timer object at each run. This is required to use it. At least, the callback function blink() is defined with a small trick at the end.

The comparison is obvious. I test the current state of the LED, if it is already switched on, I switch it off, else I switch it on. Then, I invert the state, which is the trick. I'm using the ! (not) unary operator on this Boolean variable in order to flip its value, and I assign the inverted value to the Boolean itself. I could have made this too:

```
void blink() {
  if (!currentLEDState) {
digitalWrite(ledPin, HIGH);
currentLEDState   = true ;
}

else {
digitalWrite(ledPin, LOW);
currentLEDState   = false;
}
}
```

There's really no performance gain, one way or the other. It's simply a personal decision; use whichever you prefer.

I'm personally considering the flip as a general action that has to be done every time, independent of the state. This is the reason why I proposed that you put it outside of the test structure.

Summary

This completes the first part of this book. I hope you have been able to absorb and enjoy these first (huge) steps. If not, you may want to take the time to review something you may not have clarity on; it's always worth it to better understand what you're doing.

We know a bit more about C and C++ programming, at least enough to lead us safely through the next two parts. We can now understand the basic tasks of Arduino, we can upload our firmware, and we can test them with the basic wiring.

Now, we'll move a step further into a territory where things are more practical, and less theoretical. Prepare yourself to explore new physical worlds, where you can make things talk, and communicate with each other, where your computer will be able to respond to how you feel and react, and without wires sometimes! Again, you may want to take a little time to review something you might still be a little hazy on; knowledge is power.

The future is now!

5
Sensing with Digital Inputs

Arduino boards have inputs and outputs. Indeed, this is also one of the strengths of this platform: to directly provide headers connecting the ATMega chipset legs. We can then directly wire an input or output to any other external component or circuit without having to solder.

In case you need it here, I'm reminding you of some points:

- Arduino has digital and analog inputs
- Arduino has digital outputs that can also be used to mimic analog outputs

We are going to talk about digital inputs in this chapter.

We'll learn about the global concept of sensing the world. We are going to meet a new companion named **Processing** because it is a nice way to visualize and illustrate all that we are going to do in a more graphical way. It is also a pretext to show you this very powerful and open source tool. Then, it will drive us to design the first serial communication protocol between the board and a piece of software.

We'll specifically play with switches, but we will also cover some useful hardware design patterns.

Sensing the world

In our over-connected world, a lot of systems don't even have sensors. We, humans, own a bunch of biological sensors directly in and over our body. We are able to feel temperature with our skin, light with our eyes, chemical components with both our nose and mouth, and air movement with ears. From a characteristic of our world, we are able to sense, integrate this feeling, and eventually to react.

If I go a bit further, I can remember a definition for senses from my early physiological courses at university (you remember, I was a biologist in one of my previous lives):

"Senses are physiological capacities that provide data for perception"

This basic physiological model is a nice way to understand how we can work with an Arduino board to make it sense the world.

Indeed, it introduces three elements we need:

- A capacity
- Some data
- A perception

Sensors provide new capacities

A sensor is a physical converter, able to measure a physical quantity and to translate it into a signal understandable directly or indirectly by humans or machines.

A thermometer, for example, is a sensor. It is able to measure the local temperature and to translate it into a signal. Alcohol-based or Mercury-based thermometers provide a scale written on them and the contraction/dilatation of the chemical matter according to the temperature makes them easy to read.

In order to make our Arduino able to sense the world, temperature for instance, we would have to connect a sensor.

Some types of sensors

We can find various types of types of sensors. We often think about environmental sensors when we use the term sensor.

I'll begin by quoting some environmental quantities:

- Temperature
- Humidity
- Pressure
- Gas sensors (gas-specific or not, smoke)
- Electromagnetic fields
- Anemometer (wind speed)

- Light
- Distance
- Capacitance
- Motion

This is a non-exhaustive list. For almost each quantity, we can find a sensor. Indeed, for each quantifiable physical or chemical phenomenon, there is a way to measure and track it. Each of them provides data related to the quantity measured.

Quantity is converted to data

When we use sensors, the reason is that we need to have a numeric value coming from a physical phenomenon, such as temperature or movement. If we could directly measure the temperature with our skin's thermal sensors, we would have been able to understand the relationship between the volume of chemical components and temperature itself. Because we know this relationship from other physical measures or calculations, we have been able to design thermometers.

Indeed, thermometers are converting a quantity (here a volume) related to the temperature into a value readable on the scale of the thermometer. In fact, we have a double conversion here. The volume is a function depending on the temperature. The height of the liquid inside the thermometer is a function depending on the volume of the liquid. Thus, we can understand that the height and temperature are related. This is the double conversion.

Anyway, the thermometer is a nice module that integrates all this mathematical and physical wizardry to provide data, a value: the temperature. As shown in the following figure, volume is used to provide a temperature:

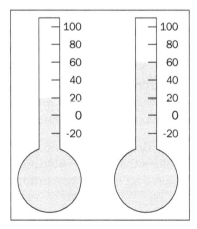

All sensors work like that. They are modules measuring physical phenomenon and providing a value. We'll see later that those values can be very different, and eventually encoded too.

Data has to be perceived

The data provided by a sensor makes more sense if it is read. This can be obvious but imagine that the reader isn't a human but is instead an instrument, a machine, or in our case, an Arduino board.

Indeed, let's take an electronic thermal sensor. At first, this one has to be supplied with electricity in order to work. Then, if we are able to supply it but unable to physically measure the electric potential generated by it from its pins, we couldn't appreciate the main value it tries to provide us: the temperature.

In our case, the Arduino would be the device that is able to convert the electric potential to something readable or at least easier to understand for us, humans. This is again a conversion. From the physical phenomenon that we want to translate, to the device displaying the value explaining the physical phenomenon, there are conversions and perceptions.

I can simplify the process as shown in the following figure:

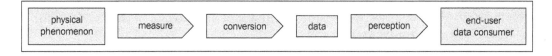

What does digital mean?

Let's define precisely what the digital term means here.

Digital and analog concepts

Digital, in the computer and electronic worlds, means discrete, which is the opposite of analog/continuous. It is also a mathematical definition. We often talk about domains to define the cases for use of digital and analog.

Usually, the analog domain is the domain related to physical measures. Our temperature can have all the values that are possible and that exist, even if our measuring equipment dosen't have an infinite resolution.

The digital domain is the one of computers. Because of the encoding and finite memory size, computers translates analog/continuous values into digital representations.

On a graph, this could be visualized as follows:

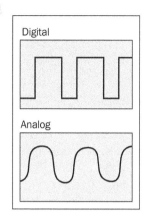

Inputs and outputs of Arduino

Arduino owns inputs and outputs. We can also distinguish analog and digital pins.

You have to remember the following points:

- Arduino provides digital pins that can be both an input or an output
- Arduino provides only analog input, not output

Inputs and outputs are pins provided by the board to communicate with external peripherals.

 Inputs provide the ability to feel the world.
Outputs provide the ability to alter the world.

We often talk about *reading pins* for inputs and *writing pins* for outputs. Indeed, from the Arduino board point of view, we are reading from the world and writing to the world, aren't we?

A digital input is a digital pin set up like an input and providing the capacity for electrical potential reading and conversion to 0 or 1 to the Arduino board. We'll illustrate this very soon using switches.

But before manipulating this directly, let me introduce a new friend named **Processing**. We'll use it to easily illustrate our Arduino tests further in the book.

Introducing a new friend – Processing

Processing is an open source programming language and Integrated Development Environment (IDE) for people who want to create images, animations, and interaction.

This major open source project was initiated in 2001 by Ben Fry and Casey Reas, two gurus and former students of John Maeda at the Aesthetics and Computation Group at the MIT Media Lab.

It is a programming framework most used by non-programmers. Indeed, it has been designed primarily for this purpose. One of the first targets of Processing is to provide an easy way of programming for non-programmers through the instant gratification of visual feedback. Indeed, as we know, programming can be very abstract. Processing natively provides a canvas on which we can draw, write, and do more. It also provides a very user-friendly IDE that we are going to see on the official website at `http://processing.org`.

You'll probably also find the term Processing written as **Proce55ing** as the domain name `processing.org` was already taken at the time of its inception.

Is Processing a language?

Processing isn't a language in the strictest sense. It's a subset of Java with some external libraries and a custom IDE.

Programming with Processing is usually performed using the native IDE comes with the download as we will see in this section.

Processing uses the Java language but provides simplified syntax and graphics programming. It also simplifies all compilations steps into a one-click action like Arduino IDE.

Like Arduino core, it provides a huge set of ready-to-use functions. You can find all references at `http://processing.org/reference`.

There is now more than one way to use Processing. Indeed, because JavaScript runtimes integrated in web browsers became more and more powerful, we can use a JavaScript derived project. You still continue to code using Java, you include this code in your webpage, and as the official website says "*Processing.js does the rest. It's not magic, but almost.*" The website is `http://processingjs.org`.

There is also something very interesting: You can package applications coded using Processing for Android mobile OS. You can read this if you are interested at `http://processing.org/learning/android`.

I will avoid going on a tangent with the JS and Android applications, but I felt it was important enough to mention these usages.

Let's install and launch it

Like the Arduino framework, the Processing framework doesn't include installation program. You just have to put it somewhere and run it from there.

The download URL is: `http://processing.org/download`.

First, download the package corresponding to your OS. Please refer to the website for the install process for your specific OS.

On OS X, you have to deflate the zip file and run the resulting file with the icon:

Processing icon

Double-click on the icon, and you'll see a pretty nice splash screen:

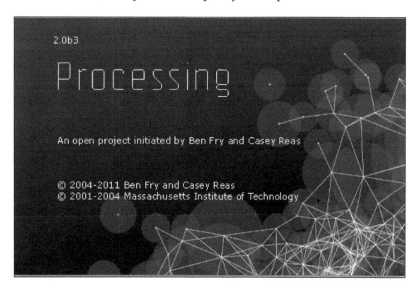

Then you'll see the Processing IDE as shown in the following image:

Processing's IDE looks like others

A very familiar IDE

Indeed, the Processing IDE looks like the Arduino one. The Processing IDE is like the father of the Arduino IDE.

This is totally normal because the Arduino IDE has been forked from the Processing IDE. Now, we are going to check that we'll be very comfortable with the Processing IDE as well.

Let's explore it and run a small example:

1. Go to **Files | Examples Basics | Arrays | ArraysObjects**.

2. Then, click on the first icon (the play symbol arrow). You should see the following screenshot:

Running ArrayObjects native example in Processing

3. Now click on the small square (stop symbol). Yes, this new playground is very familiar.

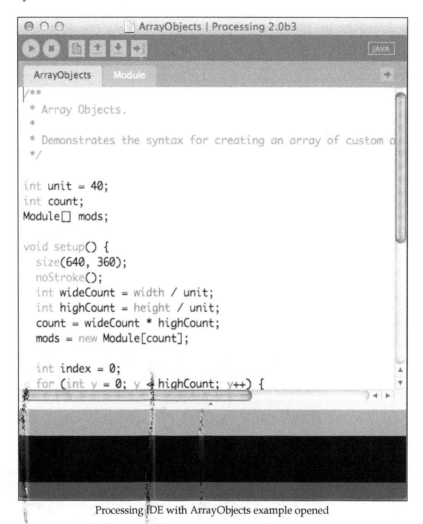

Processing IDE with ArrayObjects example opened

At the top we can see some familiar icons.

From left to right, they are as follows:

- **Run** (small arrow): This is used to compile and run your program
- **Stop** (small square): This is used to stop the program when it is running
- **New project** (small page): This is used to open a blank canvas
- **Open project** (top arrow): This is used to open an existing project

- **Save project** (down arrow): This is used to save a project
- **Export application** (right arrow): This is used to create an application

No Upload button of course. There is no need to upload anything here; we are on the computer and we only want to code applications, compile them, and run them.

With Processing, you have everything in hand to code, compile, and run.

You can have some tabs if you use more than one file in your project (especially if you use some separate Java classes).

Under this tab zone, you have the text area where you type your code. Code is colored as in the Arduino IDE, and this is very useful.

At last, at the bottom, you have the log console area where all the messages can be output, from errors to our own tracer messages.

Alternative IDEs and versioning

If you are interested in digging some IDE alternatives, I'd suggest that you use the universal and open source software development environment Eclipse. I suggest that to all the students I meet who want to go further in pure-development fields. This powerful IDE can be easily set up to support versioning.

Versioning is a very nice concept providing an easy way to track versions of your code. You can, for instance, code something, test it, back it up in your versioning system, then continue your code design. If you run it and have a nice and cute crash at some point, you can easily check the differences between your working code and the new non working one and make your troubleshooting much easier! I won't describe versioning systems in detail, but I want to introduce you to the two main systems that are widely used all over the world:

- **SVN**: http://subversion.apache.org
- **Git**: http://git-scm.com

Checking an example

Here is a small piece of code showing some cheap and easy design patterns. You can also find this code in the folder Chapter05/processingMultipleEasing/ in the code bundle:

```
// some declarations / definitions
int particlesNumber = 80;     // particles number
float[] positionsX = new float[particlesNumber]; // store particles
X-coordinates float[] positionsY = new float[particlesNumber]; //
store particles Y-coordinates
```

```
float[] radii = new float[particlesNumber];        // store particles
radii
float[] easings = new float[particlesNumber];      // store particles
easing amount

// setup is run one time at the beginning
void setup() {
  size(600, 600); // define the playground
  noStroke();      // define no stroke for all shapes drawn

  // for loop initializing easings & radii for all particles
  for (int i=0 ; i < particlesNumber ; i++)
  {
    easings[i] = 0.04 * i / particlesNumber;  // filling the easing
array
    radii[i] = 30 * i / particlesNumber ;      // filling the radii
array
  }
}

// draw is run infinitely
void draw() {
  background(34);  // define the background color of the playground

  // let's store the current mouse position
  float targetX = mouseX;
  float targetY = mouseY;

  // for loop across all particles
  for (int i=0 ; i < particlesNumber ; i++)
  {


    float dx = targetX - positionsX[i];  // calculate X distance mouse
/ particle
    if (abs(dx) > 1) {                    // if distance > 1, update
position
      positionsX[i] += dx * easings[i];
    }


    float dy = targetY - positionsY[i];   // same for Y
    if (abs(dy) > 1) {
      positionsY[i] += dy * easings[i];
    }
```

```
    // change the color of the pencil for the particle i
    fill(255 * i / particlesNumber);

    // draw the particle i
    ellipse(positionsX[i], positionsY[i], radii[i], radii[i]);
  }
}
```

You can run this piece of code. Then, you can move the mouse into the canvas and enjoy what is happening.

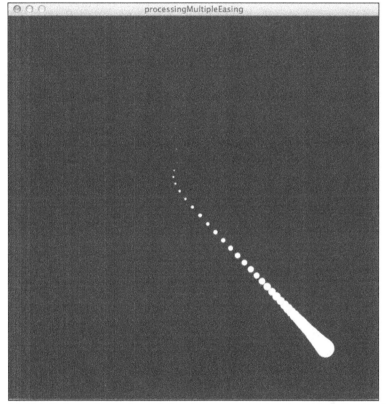

processingMultipleEasing code running and showing a strange series of particles following the mouse

First, check the code. Basically, this is Java. I guess you aren't shocked too much, are you? Indeed, Java derives from C.

You can see three main parts in your code:

- Variable declarations/definitions
- The `setup()` function that runs only once at the beginning
- The `draw()` function that runs infinitely until you press stop

Ok. You can see the `setup()` functions in the Arduino core and Processing have similar roles, and `loop()` and `draw()` too.

This piece of code shows some usual design patterns with Processing. I first initiate a variable storing the global number of particles, then I initiate some arrays for each particle I want to create. Please notice all these arrays are empty at this step!

This pattern is usual because it offers good readability and works fine too. I could have used classes or even multidimensional arrays, but in this latter case, I would not even have benefits except a shorter (but less readable) code. In all those arrays, the *N*th indexed value represents the particle *N*. In order to store/retrieve the parameters of particle *N*, I have to manipulate the *N*th value for each array. The parameters are spread inside each array but are easy to store and retrieve, aren't they?

In `setup()`, I define and instantiate the canvas and its size of 600 x 600. Then, I'm defining that there will be no stroke in any of my drawings. The stroke of a circle, for instance, is its border.

Then, I'm filling the `easing` and `radii` arrays using a `for` loop structure. This is a very usual pattern where we can use `setup()` to initialize a bunch of parameters at the beginning. Then we can check the `draw()` loop. I'm defining a color for the background. This function also erases the canvas and fills it with the color in argument. Check the background function on the reference page to understand how we can use it. This erase/fill is a nice way to erase each frame and to reset the canvas.

After this erase/fill, I'm storing the current position of the mouse for each coordinate in the local variables `targetX` and `targetY`.

The core of the program sits in the `for` loop. This loop walks over each particle and makes something for each of them. The code is quite self-explanatory. I can add here that I'm checking the distance between the mouse and each particle for each frame (each run of `draw()`), and I draw each particle by moving them a bit, according to its easing.

This is a very simple example but a nice one I used to show to illustrate the power of Processing.

Processing and Arduino

Processing and Arduino are very good friends.

Firstly, they are both open source. It is a very friendly characteristic bringing a lot of advantages like code source sharing and gigantic communities, among others. They are available for all OSes: Windows, OS X, and Linux. We also can download them for free and run them in a couple of clicks.

I began to program primarily with Processing, and I use it a lot for some of my own data visualization projects and art too. Then, we can illustrate complex and abstract data flows by smooth and primitive shapes on a screen.

What we are going to do together now is display Arduino activity on the Processing canvas. Indeed, this is a common use of Processing as an eye-friendly software for Arduino.

We are going to design a very trivial and cheap protocol of communication between the hardware and the software. This will show you the path that we'll dig further in the next chapters of this book. Indeed, if you want to get your Arduino talking with another software framework (I'm thinking about Max 6, openFrameworks, Cinder, and many others), you'll have to follow the same ways of design.

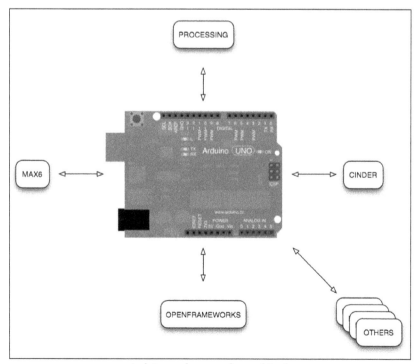

Arduino and some software friends

I often say Arduino can work as a very smart *organ* of software. If you want to connect some software to the real, physical world, Arduino is the way to go. Indeed, that way, software can sense the world, providing your computer with new features. Let's move on by displaying some physical world events on the computer.

Pushing the button

We are going to have fun. Yes, this is the very special moment when we are going to link the physical world to the virtual world. Arduino is all about this.

What is a button, a switch?

A **switch** is an electrical component that is able to break an electrical circuit. There are a lot of different types of switches.

Different types of switches

Some switches are called **toggles**. Toggles are also named continuous switches. In order to act on the circuit, the toggle can be pushed and released each time you want to act and when you release it, the action continues.

Some others are called **momentaries**. Momentaries are named **push for action** too. In order to act on the circuit, you have to push and keep the switch pushed to continue the action. If you release it, the action stops.

Usually, all our switches at home are toggles. Except the one for the mixer that you have to push to cut and release to stop it, which means it is a momentary.

A basic circuit

Here is a basic circuit with an Arduino, a momentary switch and a resistor.

We want to turn the board's built-in LED ON when we push the momentary switch and turn it OFF when we release it.

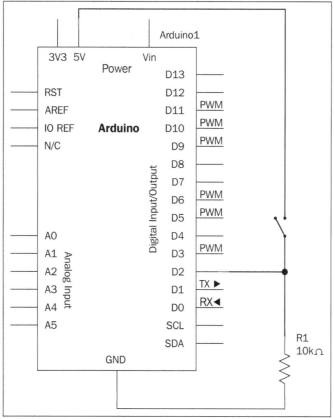

A small circuit

I'm presenting you with the circuit on which we are going to work right now. This is also a nice pretext to make you more familiar with circuit diagrams.

Wires

Each line represents a link between two components. By definition, a line is a wire and there is no electrical potential from one side to the other. It can also be defined as follows: a wire has a resistance of 0 ohm. Then we can say that two points linked by a wire have the same electrical potential.

The circuit in the real world

Of course, I didn't want to show you the next diagram directly. Now we have to build the real circuit, so please take some wires, your breadboard, and the momentary switch, and wire the whole circuit as shown in the next diagram.

You can take a resistor around 10 Kohms. We'll explain the purpose of the resistor in the next pages.

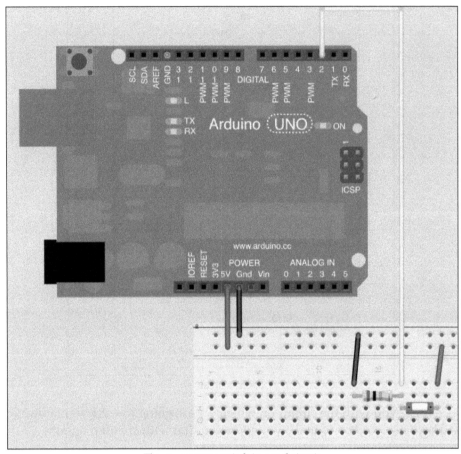

The momentary switch in a real circuit

Let's explain things a bit more.

Let's remember the breadboard wiring; I'm using cold and hot rails at the top of the breadboard (cold is blue and means ground, hot is red and means +5 V). After I have wired the ground and +5 V from the Arduino to the rails, I'm using rails to wire the other parts of the board; it is easier and requires shorter cables.

There is a resistor between the ground and the digital pin 2. There is a momentary switch between the +5 V line and the pin 2 as well. The pin 2 will be set up as an input, which means it will be able to sink current.

Usually, switches are *push-to-on*. Pushing them closes the circuit and lets the current flow. So, in that case, if I don't push the switch, there is no current from +5 V to the pin 2.

For the duration it is pressed, the circuit is closed. Then, current flows from the +5 V to the pin 2. It is a bit metaphoric and abusive, and I should say we have created an electrical potential between +5 V and the pin 2, but I need to be shorter to hit the point home.

And this resistor, why is it here?

The pull-up and pull-down concept

If the global circuit is easy, the resistor part can be a bit tricky at first sight.

A digital pin set up as an input provides the ability to *sink* current. This means it behaves like the ground. Indeed, and in fact, internally, it works exactly as if the concerned pin was connected to the ground.

With a properly coded firmware, we would have the ability to check pin 2. This means we could test it and read the value of the electrical potential. Because it is a digital input, an electrical potential near +5 V would be translated as the value HIGH, and if it is near 0 V, it will be translated as the value LOW. Both values are constants defined inside the Arduino core. But if everything seems totally perfect in a perfect digital world, it is not true.

Indeed, the input signal noise could potentially be read as a button press.

To be sure and safe, we use what we call a *pull-down resistor*. This is usually a high impedance resistance that provides a current sink to the digital pin considered, making it safer at the value 0 V if the switch is not pressed. Pull down to be more consistently recognized as a LOW value, pull up to be more consistently recognized as the HIGH value.

Of course, the global energy consumption increases a bit. In our case, this is not important here but you have to know that. On this same concept, a pull-up resistor can be used to link the +5 V to the digital output. Generally, you should know that a chipset's I/O shouldn't be floating.

Here is what you have to remember:

Type of Digital Pin	Input	Output
Pull Resistor	Pull-down resistor	Pull-up resistor

Let's design a firmware

We want to push a switch, and particularly, this action has to turn the LED ON. We are going to write a pseudocode first.

The pseudocode

Here is a possible pseudocode. Following are the steps we want our firmware to follow:

1. Define the pins.
2. Define a variable for the current switch state.
3. Set up the LED pin as an output.
4. Set up the switch pin as an input.
5. Set up an infinite loop. In the infinite loop do the following:

    1. Read the input state and store it.
    2. If the input state is HIGH, turn the LED ON.
    3. Else turn the LED OFF.

The code

Here is a translation of this pseudocode in valid C code:

```
const int switchPin = 2;     // pin of the digital input related to
the switch
const int ledPin =  13;      // pin of the board built-in LED

int switchState = 0;         // storage variable for current switch
state

void setup() {
  pinMode(ledPin, OUTPUT);   // the led pin is setup as an output
  pinMode(switchPin, INPUT); // the switch pin is setup as an input
}

void loop(){
```

```
   switchState = digitalRead(switchPin);   // read the state of the
digital pin 2

   if (switchState == HIGH) {      // test if the switch is pushed or
not

      digitalWrite(ledPin, HIGH);  // turn the LED ON if it is currently
pushed
   }
   else {
      digitalWrite(ledPin, LOW);   // turn the LED OFF if it is
currently pushed
   }
}
```

As usual you can also find the code in the `Chapter05/MonoSwitch/` folder available for download along with other code files on Packt Publishing's site.

Upload it and see what happens. You should have a nice system on which you can push a switch and turn on an LED. Splendid!

Now let's make the Arduino board and Processing communicate with each other.

Making Arduino and Processing talk

Let's say we want to visualize our switch's manipulations on the computer.

We have to define a small communication protocol between Arduino and Processing. Of course, we'll use a serial communication protocol because it is quite easy to set it up and it is light.

We could design a protocol as a library of communication. We only design a protocol using the native Arduino core at the moment. Then, later in this book, we will design a library.

The communication protocol

A communication protocol is a system of rules and formats designed for exchanging messages between two entities. Those entities can be humans, computers and maybe more.

Indeed, I'd use a basic analogy with our language. In order to understand each other, we have to follow some rules:

- Syntactic and grammatical rules (I have to use words that you know)
- Physical rules (I have to talk loud enough)
- Social rules (I shouldn't insult you just before asking you for the time)

I could quote many other rules like the speed of talking, the distance between the two entities, and so on. If each rule is agreed upon and verified, we can talk together. Before designing a protocol, we have to define our requirements.

Protocol requirements

What do we want to do?

We need a communication protocol between our Arduino and Processing inside the computer. Right! These requirements are usually the same for a lot of communication protocols you'll design.

Here is a short list of very important ones:

- The protocol must be expandable without having to rewrite everything each time I want to add new message types
- The protocol must be able to send enough data quite quickly
- The protocol must be easy to understand and well commented, especially for open source and collaborative projects

Protocol design

Each message will be 2 bytes in size. This is a common data packet size and I propose to organize data like this:

- **Byte 1**: switch number
- **Byte 2**: switch state

The fact that I defined byte 1 as a representation of the switch number is typically because of the requirement of expandability. With one switch, the number will be 0.

I can easily instantiate serial communication between the board and the computer. Indeed, we already made that when we used Serial Monitoring at least on the Arduino side.

How can we do that using Processing?

The Processing code

Processing comes with very useful set of libraries already integrated into its core. Specifically, we are going to use the serial library.

Let's sketch a pseudocode first, as usual.

Sketching a pseudocode

What do we want the program to do?

I propose to have a big circle. Its color will represent the switch's state. *Dark* will mean released, and *green* will mean pushed.

The pseudocode can be created as follows:

1. Define and instantiate the serial port.
2. Define a current drawing color to dark.
3. In the infinite loop, do the following:

    1. Check if the serial port and grab data have been received.
    2. If data indicates that state is off, change current drawing from color to dark.
    3. Else, change current drawing color to green.
    4. Draw the circle with the current drawing color.

Let's write that code

Let's open a new processing canvas.

Because the Processing IDE works like the Arduino IDE and needs to create all saved project files in a folder, I'd suggest that you directly save the canvas, even empty, in the right place on your disk. Call it `processingOneButtonDisplay`.

You can find the code in the `Chapter05/processingOneButtonDisplay/` folder available for download along with other code files on Packt's site.

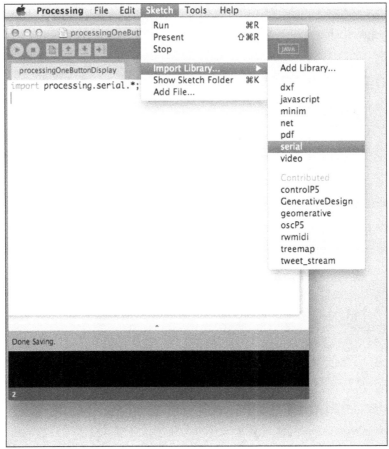

Making a library inclusion in your code

To include the serial library from the Processing core, you can go to **Sketch | Import Library... | serial**. It adds this row to your code: `processing.serial.*;`

You could also type this statement by yourself.

Following is the code, with a lot of comments:

```
import processing.serial.*;

Serial theSerialPort;              // create the serial port object
int[] serialBytesArray = new int[2];  // array storing current message
int switchState;                   // current switch state
int switchID;                      // index of the switch
```

```
int bytesCount = 0;              // current number of bytes relative
to messages
boolean init = false;            // init state
int fillColor = 40;              // defining the initial fill color


void setup(){

  // define some canvas and drawing parameters
  size(500,500);
  background(70);
  noStroke();

  // printing the list of all serial devices (debug purpose)
  println(Serial.list());

  // On osx, the Arduino port is the first into the list
  String thePortName = Serial.list()[0];

  // Instantate the Serial Communication
  theSerialPort = new Serial(this, thePortName, 9600);
}


void draw(){

  // set the fill color
  fill(fillColor);

  // draw a circle in the middle of the screen
  ellipse(width/2, height/2, 230, 230);
}



void serialEvent(Serial myPort) {

  // read a byte from the serial port
  int inByte - myPort.read();

  if (init == false) {           // if there wasn't the first hello
    if (inByte == 'Z') {         // if the byte read is Z
      myPort.clear();            // clear the serial port buffer
      init = true;               // store the fact we had the first
hello
      myPort.write('Z');         // tell the Arduino to send more !
    }
```

```
    }
    else {                              // if there already was the first hello

       // Add the latest byte from the serial port to array
       serialBytesArray[bytesCount] = inByte;
       bytesCount++;

       // if the messages is 2 bytes length
       if (bytesCount > 1 ) {
          switchID = serialBytesArray[0]; // store the ID of the switch
          switchState = serialBytesArray[1]; // store the state of the
   switch

          // print the values (for debugging purposes):
          println(switchID + "\t" + switchState);

          // alter the fill color according to the message received from
   Arduino
          if (switchState == 0) fillColor = 40;
          else fillColor = 255;


          // Send a capital Z to request new sensor readings
          myPort.write('Z');

          // Reset bytesCount:
          bytesCount = 0;
       }
    }
}
```

Variable definitions

theSerialPort is an object of the Serial library. I have to create it first.

serialBytesArray is an array of two integers used to store messages coming from Arduino. Do you remember? When we designed the protocol, we talked about 2 byte messages.

switchState and switchID are global but temporary variables used to store the switch state and the switch ID corresponding to the message coming from the board. Switch ID has been put there for (close) future implementation to distinguish the different switches in case we use more than one.

`bytesCount` is a useful variable tracking the current position in our message reading.

`init` is defined to `false` at the beginning and becomes `true` when the first byte from the Arduino (and a special one, z) has been received for the first time. It is a kind of first-contact purpose.

Then, we keep a trace of the fill color and the initial one is 40. 40 is only an integer and will be used a bit further as an argument of the function `fill()`.

setup()

We define the canvas (size, background color, and no stroke).

We print a list of all the serial ports available on your computer. This is debug information for the next statement where we store the name of the first serial port into a String. Indeed, you could be led to change the array element from 0 to the correct one according to the position of your Arduino's port in the printed list.

This String is then used in the very important statement that instantiates serial communication at 9600 bauds.

This `setup()` function, of course, runs only once.

draw()

The draw function is very light here.

We pass the variable `fillColor` to the `fill()` function, setting up the color with which all further shapes will be filled.

Then, we draw the circle with the ellipse function. This function takes four arguments:

- x coordinates of the center of the ellipse (here `width/2`)
- y coordinates of the center of the ellipse (here `height/2`)
- Width of the ellipse (here `230`)
- Height of the ellipse (here `230`)

`width` and `height` colored in blue in the Processing IDE are the current width and height of the canvas. It is very useful to use them because if you change the `setup()` statement by choosing a new size for the canvas, all `width` and `height` in your code will be updated automatically without needing to change them all manually.

Please keep in mind that an ellipse with same values for `width` and `height` is a circle (!). Ok. But where is the magic here? It will only draw a circle, every time the same one (size and position). `fillColor` is the only variable of the `draw()` function. Let's see that strange callback named `serialEvent()`.

The serialEvent() callback

We talked about callbacks in *Chapter 4, Improve Programming with Functions, Math, and Timing.*

Here, we have a pure callback method in Processing. This is an event-driven callback. It is useful and efficient not to have to poll every time our serial port wants to know if there is something to read. Indeed, user interfaces related events are totally less numerous than the number of Arduino board's processor cycles. It is smarter to implement a callback in that case; as soon as a serial event occurs (that is, a message is received), we execute a series of statements.

`myPort.read()` will first read the bytes received. Then we make the test with the `init` variable. Indeed, if this is the very first message, we want to check if the communication has already begun.

In the case where it is the first hello (`init == false`), if the message coming from the Arduino Board is `z`, Processing program clear its own serial port, stores the fact the communication has just started, and resends back `z` to the Arduino board. It is not so tricky.

It can be illustrated as follows:

Imagine we can begin to talk only if we begin by saying "hello" to each other. We aren't watching each other (no event). Then I begin to talk. You turn your head to me (serial event occurs) and listen. Am I saying "hello" to you? (whether the message is `z`?). If I'm not, you just turn your head back (no `else` statement). If I am, you answer "hello" (sending back `z`) and the communication begins.

What happens then?

If communication has already begun, we have to store bytes read into the `serialBytesArray` and increment the `bytesCount`. While bytes are being received and `bytesCount` is smaller or equal to 1, this means we don't have a complete message (a complete message is two bytes) and we store more bytes in the array.

As soon as the bytes count equals 2, we have a complete message and we can "split" it into the variables `switchID` and `switchState`. Here's how we do that:

```
switchID = serialBytesArray[0];
switchState = serialBytesArray[1];
```

This next statement is a debug one: we print each variable. Then, the core of the method is the test of the switchState variable. If it is 0, it means the switch is released, and we modify the fillColor to 40 (dark color, 40 means the value 40 for each RGB component; check color() method in Processing reference at http:// processing.org/reference/color_.html). If it isn't 0, we modify the fillColor to 255, which means white. We could be a bit safer by not using only else, but else if (switchState ==1) also.

Why? Because if we are not sure about all the messages that can be sent (lack of documentation or whatever else making us unsure), we can modify the color to white *only* if switchState equals 1. This concept can be done at the optimization state too, but here, it is quite light so we can leave it like that.

Ok. It is a nice, heavy piece, right? Now, let's see how we have to modify the Arduino code. Do you remember? It isn't communication ready yet.

The new Arduino firmware talk-ready

Because we now have a nice way to display our switch state, I'll remove all things related to the built-in LED of the board and following is the result:

```
const int switchPin = 2;      // pin of the digital input related to
the switch
int switchState = 0;          // storage variable for current switch
state

void setup() {
  pinMode(switchPin, INPUT); // the switch pin is setup as an input
}

void loop(){
    switchState = digitalRead(switchPin);
}
```

What do we have to add? All the Serial stuff. I also want to add a small function dedicated to the first "hello".

Here is the result, then we will see the explanations:

```
const int switchPin = 2;      // pin of the digital input related to
the switch
int switchState = 0;          // storage variable for current switch
state
int inByte = 0;

void setup() {
```

```
    Serial.begin(9600);
    pinMode(switchPin, INPUT); // the switch pin is setup as an input
    sayHello();
}

void loop(){

    // if a valid byte is received from processing, read the digital in.
    if (Serial.available() > 0) {
      // get incoming byte:
      inByte = Serial.read();
      switchState = digitalRead(switchPin);

      // send switch state to Arduino
      Serial.write("0");
      Serial.write(switchState);
    }
}

void sayHello() {
   while (Serial.available() <= 0) {
     Serial.print('Z');    // send a capital Z to Arduino to say
"HELLO!"
     delay(200);
   }
}
```

I'm defining one new variable first: inByte. This stores the bytes read. Then inside the setup() method, I'm instantiating serial communication as we already learned to do with Arduino. I'm setting up the pinMode method of the switch pin then, I'm calling sayHello().

This function just waits for something. Please focus on this.

I'm calling this function in `setup()`. This is a *simple* call, not a callback or whatever else. This function contains a `while` loop while `Serial.available()` is smaller or equal to zero. What does this mean? It means this function pauses the `setup()` method while the first byte comes to the serial port of the Arduino board. The `loop()` done doesn't run while the `setup()` done has finished, so this is a nice trick to wait for the first external event; in this case, the first communication. Indeed, the board is sending the message `z` (that is, the "hello") while Processing doesn't answer.

The consequence is that when you can plug in your board, it sends `z` continuously while you run your Processing program. Then the communication begins and you can push the switch and see what is happening. Indeed, as soon as the communication begins, `loop()` begins its infinite loop. At first a test is made at each cycle and we only test if a byte is being received. Whatever the byte received (Processing only sends `z` to the board), we read the digital pin of the switch and send back two bytes. Here too, pay attention please: each byte is written to the serial port using `Serial.write()`. You have to send 2 bytes, so you stack two `Serial.write()`. The first byte is the number (ID) of the switch that is pushed/released; here, it is not a variable because we have one and only one switch, so it is an integer 0. The second byte is the switch state. We just saw here a nice design pattern involving the board, an external program running on a computer and a communication between both of them.

Now, let's go further and play with more than one switch.

Playing with multiple buttons

We can extrapolate our previously designed logic with more than one switch.

There are many ways to use multiple switches, and, in general, multiple inputs on the Arduino. We're going to see a cheap and easy first way right now. This way doesn't involve multiplexing a lot of inputs on only a couple of Arduino inputs but a basic one to one wiring where each switch is wired to one input. We'll learn multiplexing a bit later (in the next chapter).

The circuit

Following is the circuit diagram required to work with multiple switches:

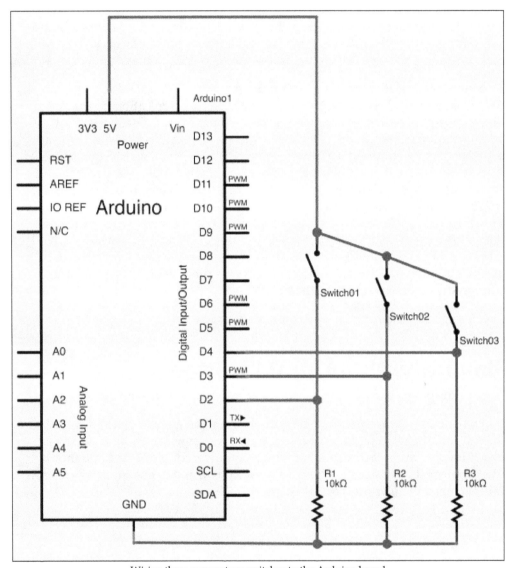

Wiring three momentary switches to the Arduino board

The schematic is an extrapolation of the previous one that showed only one switch. We can see the three switches between the +5 V and the three pull-down resistors. Then we can also see the three wires going to digital inputs 2 to 4 again.

Here is a small memory refresh: Why didn't I use the digital pins 0 or 1?

Because I'm using serial communication from the Arduino, we cannot use the digital pins 0 and 1 (each one respectively corresponding to RX and TX used in serial communication). Even if we are using the USB link as the physical support for our serial messages, the Arduino board is designed like that and we have to be very careful with it.

Here is the circuit view with the breadboard. I voluntarily didn't align every wire. Why? Don't you remember that I want you to be totally autonomous after reading this book and yes, you'll find many schematics in the real world made sometimes like that. You have to become familiar with them too. It could be an (easy) homework assignment.

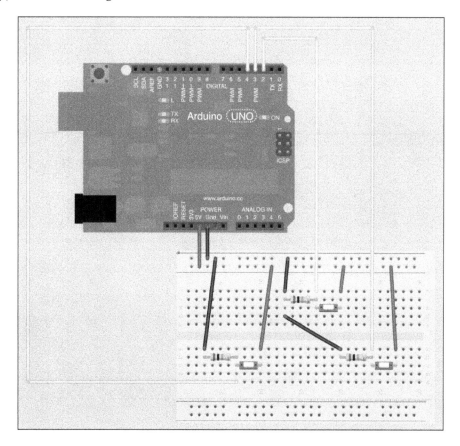

The preceding circuit shows the three switches, the three pull-down resistors, and the Arduino board.

Both source codes have to be modified to provide a support for the new circuit.

Let's add things there.

The Arduino code

Here is the new code; of course, you can find it in the Chapter05/
MultipleSwitchesWithProcessing/ folder available for download
along with other code files on Packt's site:

```
#define switchesNumber 3              // define the number of switches
as a constant

int switchesStates[switchesNumber] ; // array storing current switches
states
int inByte = 0;

void setup() {
  Serial.begin(9600);

  // initiating each pins as input and filling switchesStates with
zeroes
  for(int i = 0 ; i < switchesNumber ; i++)
  {
// BE CAREFUL TO THAT INDEX
pinMode(i + 2, INPUT); // the switch pin is setup as an input

    switchesStates[i] = 0 ;
  }

  sayHello(); // waiting for the processing program hello answer
}

void loop(){

  // if a valid byte is received from processing, read all digital
inputs.
  if (Serial.available() > 0) {

    // get incoming byte
    inByte = Serial.read();

    for(int i = 0 ; i < switchesNumber ; i++)
    {
      switchesStates[i] = digitalRead(i+2); // BE CAREFUL TO THAT
INDEX
    // WE ARE STARTING FROM PIN 2 !
```

```
      Serial.write(i);                    // 1st byte = switch number (0
  to 2)
      Serial.write(switchesStates[i]); // 2nd byte = the switch i
  state
    }
  }
}

void sayHello() {
  while (Serial.available() <= 0) {
    Serial.print('Z');    // send a capital Z to Arduino to say
  "HELLO!"
    delay(200);
  }
}
```

Let's explain this code.

At first, I defined a constant switchesNumber to the number 3. This number can be changed to any other number from 1 to 12. This number represents the current number of switches wired to the board from digital pin 2 to digital pin 14. All switches have to be linked without an empty pin between them.

Then, I defined an array to store the switch's states. I declared it using the switchesNumber constant as the length. I have to fill this array with zeroes in the setup() method, that I made with a for loop. It provides a safe way to be sure that all switches have a release state in the code.

I still use the sayHello() function, to set up the communication start with Processing.

Indeed, I have to fill each switch state in the array switchesStates so I added the for loop. Please notice the index trick in each for loop. Indeed, because it seems to be more convenient to start from 0 and because in the real world we mustn't use digital pins 0 and 1 while using serial communications, I added 2 as soon as I dealt with the real number of the digital pin, that is, with the two functions pinMode() and digitalRead().

Now, let's upgrade the Processing code too.

The Processing code

Here is the new code; you can find it in the `Chapter05/`
`MultipleSwitchesWithProcessing/` folder available for download along with
other code files on Packt's site:

```
import processing.serial.*;

int switchesNumber = 2;

Serial theSerialPort;                    // create the serial port object
int[] serialBytesArray = new int[2];  // array storing current message
int switchID;                            // index of the switch
int[] switchesStates = new int[switchesNumber]; // current switch
state
int bytesCount = 0; // current number of bytes relative to messages
boolean init = false;                    // init state
int fillColor = 40;                      // defining the initial fill
color

// circles display stuff
int distanceCircles ;
int radii;

void setup() {

  // define some canvas and drawing parameters
  size(500, 500);
  background(70);
  noStroke();
  distanceCircles = width / switchesNumber;
  radii = distanceCircles/2;

  // printing the list of all serial devices (debug purpose)
  println(Serial.list());

  // On osx, the Arduino port is the first into the list
  String thePortName = Serial.list()[0];

  // Instantate the Serial Communication
  theSerialPort = new Serial(this, thePortName, 9600);

  for (int i = 0 ; i < switchesNumber ; i++)
  {
    switchesStates[i] = 0;
```

```
    }
}

void draw() {

  for (int i = 0 ; i < switchesNumber ; i++)
  {
    if (switchesStates[i] == 0) fill(0);
    else fill(255);

    // draw a circle in the middle of the screen
    ellipse(distanceCircles * (i + 1) - radii, height/2, radii,
radii);
  }
}


void serialEvent(Serial myPort) {

  // read a byte from the serial port
  int inByte = myPort.read();

  if (init == false) {         // if this is the first hello
    if (inByte == 'Z') {       // if the byte read is Z
      myPort.clear();          // clear the serial port buffer
      init = true;             // store the fact we had the first
hello
      myPort.write('Z');       // tell the Arduino to send more !
    }
  }
  else {                       // if there already was the first hello

    // Add the latest byte from the serial port to array
    serialBytesArray[bytesCount] = inByte;
    bytesCount++;

    // if the messages is 2 bytes length
    if (bytesCount > 1 ) {
      switchID = serialBytesArray[0];      // store the ID of the
switch
      switchesStates[switchID] = serialBytesArray[1]; // store state
of the switch

      // print the values (for debugging purposes):
      println(switchID + "\t" + switchesStates[switchID]);
```

```
        // Send a capital Z to request new sensor readings
        myPort.write('Z');

        // Reset bytesCount:
        bytesCount = 0;
      }
    }
  }
```

Following is a screenshot of the render of this code used with five switches while I was pushing on the fourth button:

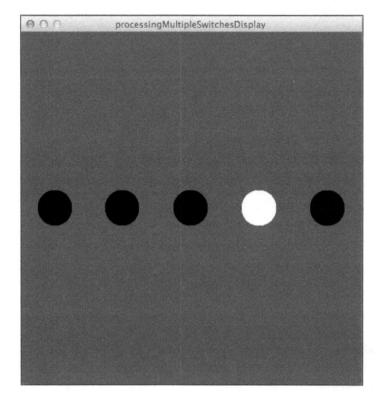

So, what did I alter?

Following the same concept as with the Arduino code, I added a variable (not a constant here), named switchesNumber. A nice evolution could be to add something to the protocol about the number of the switch. For instance, the Arduino board could inform Processing about the switch's number according to only one constant defined in the Arduino firmware. This would save the manual update of the processing code when we change this number.

I also transformed the variable `switchState` into an array of integers `switchesStates`. This one stores all the switches' states. I added two variables related to the display: `distanceCircles` and `radii`. Those are used for dynamically displaying the position of circles according to the number of switches. Indeed, we want one circle per switch.

The `setup()` function is almost the same as before.

I'm calculating here the distance between two circles by dividing the width of the canvas by the number of circles. Then, I'm calculating the radii of each circle by using the distance between them divided by 2. These numbers can be changed. You could have a very different aesthetical choice.

Then the big difference here is also the `for` loop. I'm filling the whole `switchesStates` array with zeroes to initialize it. At the beginning, none of the switches are pushed. The `draw()` function now also includes a `for` loop. Pay attention here. I removed the `fillColor` method because I moved the fill color choice to the draw. This is an alternative, showing you the flexibility of the code.

In the same for loop, I'm drawing the circle number *i*. I will let you check for yourself how I have placed the circles. The `serialEvent()` method doesn't change a lot either. I removed the fill color change as I wrote before. I also used the `switchesStates` array, and the index provided by the first byte of the message that I stored in `switchID`.

Now, you can run the code on each side after you have uploaded the firmware on the Arduino board.

Magic? I guess you now know that it isn't magic at all, but beautiful, maybe.

Let's go a bit further talking about something important about switches, but also related to other switches.

Understanding the debounce concept

Now here is a small section that is quite cool and light compared to analog inputs, which we will dive into in the next chapter.

We are going to talk about something that happens when someone pushes a button.

What? Who is bouncing?

Now, we have to take our microscopic biocybernetic eyes to zoom into the switch's structure.

A switch is made with pieces of metal and plastic. When you push the cap, a piece of metal moves and comes into contact with another piece of metal, closing the circuit. Microscopically and during a very small time interval, things aren't that clean. Indeed, the moving piece of metal bounces against the other part. With an oscilloscope measuring the electrical potential at the digital pin of the Arduino, we can see some noise in the voltage curve around 1 ms after the push.

These oscillations could generate incorrect inputs in some programs. Imagine, that you want to count the states transitions in order, for instance, to run something when the user pushed the switch seven times. If you have a bouncing system, by pushing only once, the program could count a lot of transitions even if the user pushed the switch only once.

Check the next graph. It represents the voltage in relation to time. The small arrows on the time axis show the moment when the switch has been pushed:

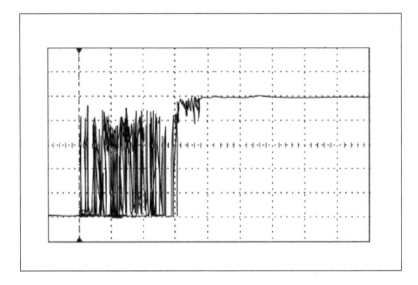

How can we deal with these oscillations?

How to debounce

We have two distinct elements on which we can act:

- The circuit itself
- The firmware

The circuit itself can be modified. I could quote some solutions such as adding diodes, capacitors, and some Schmitt trigger inverters. I won't explain that solution in detail because we are going to do that in software, but I can explain the global concept. The capacitor in that case will be charged and discharged while the switch will be bouncing, smoothing those peaks of noise. Of course, some tests are needed in order to find the perfect components fitting your precise needs.

The firmware can also be modified.

Basically, we can use a time-based filter, because the bounce occurs during a particular amount of time.

Following is the code, then will come explanations:

```
const int switchPin = 2;       // pin of the digital input related to
the switch
const int ledPin =  13;        // pin of the board built-in LED

int switchState = 0;           // storage variable for current switch
state
int lastSwitchState= LOW;

// variables related to the debouncing system
long lastDebounceTime = 0;
long debounceDelay = 50;

void setup() {
  pinMode(ledPin, OUTPUT);    // the led pin is setup as an output
  pinMode(switchPin, INPUT); // the switch pin is setup as an input
}

void loop(){

  // read the state of the digital pin
  int readInput = digitalRead(switchPin);

  // if freshly read state is different than the last debounced value
  if (readInput != lastSwitchState){
   // reset the debounce counter by storing the current uptime ms
   lastDebounceTime = millis();
  }

  // if the time since the last debounce is greater than the debounce
delay
  if ( (millis() - lastDebounceTime) > debounceDelay ){
```

```
    // store the value because it is a debounced one and we are safe
    switchState = readInput;
    }

    // store the last read state for the next loop comparison purpose
    lastSwitchState = readInput;

    // modify the LED state according to the switch state
    if (switchState == HIGH)
    {       // test if the switch is pushed or not

      digitalWrite(ledPin, HIGH);   // turn the LED ON if it is currently
    pushed
      }
      else
      {
      digitalWrite(ledPin, LOW);    // turn the LED OFF if it is
    currently pushed
      }
    }
```

Following is an example of the debouncing cycle.

At the beginning, I defined some variables:

- lastSwitchState: This stores the last read state
- lastDebounceTime: This stores the moment when the last debounce occurred
- debounceDelay: This is the value during which nothing is taken as a safe value

We are using millis() here in order to measure the time. We already talked about this time function in *Chapter 4, Improve Programming with Functions, Math, and Timing*.

Then, at each loop() cycle, I read the input but basically I don't store it in the switchState variable that is used to the test to turning ON or OFF the LED. Basically, I used to say that switchState is the official variable that I don't want to modify before the debounce process. Using other terms, I can say that I'm storing something in switchState only when I'm sure about the state, not before.

So I read the input at each cycle and I store it in readInput. I compare readInput to the lastSwitchState variable that is the last read value. If both variables are different, what does it mean? It means a change occurs, but it can be a bounce (unwanted event) or a real push. Anyway, in that case, we reset the counter by putting the current time provided by millis() to lastDebounceTime.

Then, we check if the time since the last debounce is greater than our delay. If it is, then we can consider the last `readInput` in this cycle as the real switch state and we can store it into the corresponding variable. In the other case, we store the last read value into `lastSwitchState` to keep it for the next cycle comparison.

This method is a general concept used to smooth inputs.

We can find here and there some examples of software debouncing used not only for switches but also for noisy inputs. In everything related to a user-driven event, I would advise using this kind of debouncer. But for everything related to system communication, debounce can be very useless and even a problem, because we can ignore some important messages and data. Why? Because a communication system is much faster than any user, and if we can use 50 ms as the time during which nothing is considered as a real push or a real release with users, we cannot do that for very fast chipset signals and other events that could occurs between systems themselves.

Summary

We have learnt a bit more about digital inputs. Digital inputs can be used *directly*, as we just did, or also *indirectly*. I'm using this term because indeed, we can use other peripherals for encoding data before sending them to digital inputs. I used some distance sensors that worked like that, using digital inputs and not analog inputs. They encoded distance and popped it out using the I2C protocol. Some specific operations were required to extract and use the distance. In this way, we are making an indirect use of digital inputs.

Another nice way to sense the world is the use of analog inputs. Indeed, this opens a new world of continuous values. Let's move on.

6
Sensing the World – Feeling with Analog Inputs

The real world isn't digital. My digital-art-based vision shows me *The Matrix* behind things and huge digital waterfalls between things. In this chapter, however, I need to convey to you the relationship between digital and analog, and we need to understand it well.

This chapter is a good one but a huge one. Don't be afraid. We'll also discuss new concepts a lot while writing and designing pure C++ code.

We are going to describe together what analog inputs are. I'm also going to introduce you to a new and powerful friend worthy of respect, Max 6 framework. Indeed, it will help us a bit like Processing did – to communicate with the Arduino board. You'll realize how important this is for computers, especially when they have to sense the world. A computer with the Max 6 framework is very powerful, but a computer with the Max 6 framework and the Arduino plugin can feel much characteristics of the physical world, such as pressure, temperature, light, color, and many more. Arduino, as we have already seen, behaves a bit like a very powerful organ able to…*feel*.

If you like this concept of feeling things, and especially that of making other things react to these feelings, you'll love this chapter

Sensing analog inputs and continuous values

There's no better way to define analog than by comparing it to digital. We just talked about digital inputs in the previous chapter, and you now know well about the only two values those kind of inputs can read. It is a bit exhausting to write it, and I apologize because this is indeed more a processor constraint than a pure input limitation. By the way, the result is that a digital input can only provide 0 or 1 to our executed binary firmware.

Analog works totally differently. Indeed, analog inputs can continuously provide variable values by measuring voltage from 0 V to 5 V. It means a value of 1.4 V and another value of 4.9 V would be interpreted as totally different values. This is very different from a digital input that could interpret them as...1. Indeed, as we already saw, a voltage value greater than 0 is usually understood as 1 by digital inputs. 0 is understood as 0, but 1.4 would be understood as 1; this we can understand as HIGH, the ON value, as opposed to the OFF, which comes from the 0 V measure.

Here, in the continuous world of analog inputs, we can sense a flow between the different values, where digital inputs can provide only steps. This is one of the reasons why I'm always using the term "feeling". Yes, when you can measure a lot of values, this is near to sensing and feeling. This is a bit of humanization of the electronic hardware, and I totally assume that.

How many values can we distinguish?

The term "a lot" isn't precise. Even if we are in a new continuous field of measure, we are still in the digital world, the one of the computers. So how many values can be distinguished by Arduino's analog inputs? 1024.

Why 1024? The reason is easy to understand if you understand how Arduino can feel continuous values.

Because Arduino's chip works in the digital domain for all calculations, we have to convert analog values from 0 V to 5 V to a digital one. The purpose of the **analog-to-digital converter**, housed within the chipset itself, is exactly this. This device is also referred to using the acronym ADC.

Arduino's ADCs have a 10-bit resolution. This means that every analog value is encoded and mapped to a 10-bit, encoded integer value. The maximum number encodable using this encoding system is 1111111111 in the binary system, which means 1023 in the decimal system. If I consider the first number to be 0, we have a total of 1024 values represented. A 1024-value resolution provides a very comfortable field of sensing as we are going to see in the next few pages.

Let's see how we can use these precious inputs with Arduino.

Reading analog inputs

Because we are now more familiar with circuits and code, we can work with a small project while still explaining concepts. I'm going to describe a simple example of circuits and code using a **potentiometer** only.

The real purpose of the potentiometer

First, let's grab a potentiometer. A potentiometer is, if you remember correctly from the first chapter of this book, a variable resistor.

Considering Ohm's law, which links voltage, current, and resistance value, we can understand that, for a constant current, we can make the voltage vary by changing the value of the resistance of the potentiometer. Indeed, because some of us haven't dusted off our elementary electronics course textbook in many years, how about a refresher? Here's Ohm's law:

$V = R * I$

Here, V is the voltage in Volts, R the resistance in Ohms, and I the current in Amperes.

So now, to define the purpose of a potentiometer:

 The potentiometer is your way to change continuously a variable in your running code from the physical world.

 Always remember:
Use 10-bit resolution, and you'll be the master of analog inputs!

Changing the blinking delay of an LED with a potentiometer

The following figure is the most basic circuit to illustrate the concept of analog inputs with the Arduino board:

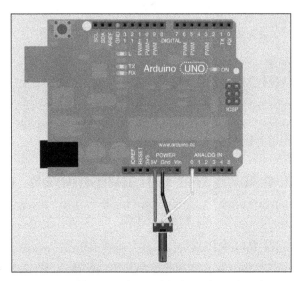

A potentiometer connected to the Arduino board

Check the corresponding electrical diagram for connections:

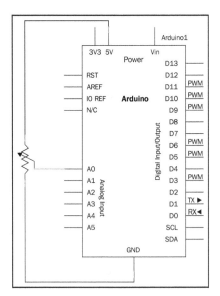

Analog input 0 is measuring the voltage

Now let's see the code we have to use.

Like the function `digitalRead()`, which can read the value of digital inputs on the Arduino, there is `analogRead()` for doing the same with analog inputs.

The intention here is to read the value as a pause value in our program for the purpose of controlling the blink rate of an LED. In code, we'll be using the `delay()` function.

Here's an example:

```
int potPin = 0;      // pin number where the potentiometer is connected
int ledPin = 13 ;    // pin number of the on-board LED
int potValue = 0 ;   // variable storing the voltage value measured at
potPin pin

void setup() {
  pinMode(ledPin, OUTPUT);   // define ledPin pin as an output
}

void loop(){
  potValue = analogRead(potPin); // read and store the read value at
potPin pin

  digitalWrite(ledPin, HIGH);    // turn on the LED
  delay(potValue);               // pause the program during potValue
millisecond
  digitalWrite(ledPin, LOW);     // turn off the LED
  delay(potValue);               // pause the program during potValue
millisecond
}
```

Upload the code. Then turn the pot a bit, and observe the output.

After the variable definition, I'm defining the `ledPin` pin as output in the `setup()` function in order to be able to drive current to this pin. Actually, I'm using pin 13 in order to simplify our tests. Don't forget pin 13 is the surface-mounted LED on the Arduino board.

Then, the magic happens in the `loop()` function.

I'm first reading the value at the `potPin` pin. As we discussed before, the value returned by this function is an integer between 0 and 1023. I'm storing it in the `potValue` variable to keep the LED ON, but also to keep it OFF.

Then, I'm turning OFF and ON the LED with some delay between status changes. The smart thing here is to use `potValue` as the delay. Turned on one side completely, the potentiometer provides a value of 0. Turned on the other side completely, it provides 1023, which is a reasonable and user-friendly delay value in milliseconds.

The higher the value is, the longer the delay.

In order to be sure you understood the physical part of this, I'd like to explain a bit more about voltage.

The +5 V and ground pins of the Arduino supply the potentiometer the voltage. Its third leg provides a way to vary the voltage by varying the resistance. The Arduino's analog inputs are able to read this voltage. Please notice that analog pins on the Arduino are inputs only. This is also why, with analog pins, we don't have to worry about precision in the code like we have for digital pins.

So let's modify the code a bit in order to read a voltage value.

How to turn the Arduino into a low voltage voltmeter?

Measuring voltage requires two different points on a circuit. Indeed, a voltage is an electrical potential. Here, we have (only) that analog pin involved in our circuit to measure voltage. What's that ?!

Simple! We're using the +5 V supply from Vcc as a reference. We control the resistance provided by the potentiometer and supply the voltage from the Vcc pin to have something to demonstrate.

If we want to use it as a real potentiometer, we have to supply another part of a circuit with Vcc too, and then connect our A0 pin to another point of the circuit.

As we saw, the `analogRead()` function only provides integers from 0 to 1023. How can we have real electrical measures displayed somewhere?

Here's how it works:

The range 0 to 1023 is mapped to 0 to 5V. That comes built into the Arduino. We can then calculate the voltage as follows:

V = 5 * (analogRead() value / 1023)

Let's implement it and display it on our computer by using the serial monitor of the Arduino IDE:

```
int potPin = 0;      // pin number where the potentiometer is connected
int ledPin = 13 ;    // pin number of the on-board LED
int potValue = 0 ;   // variable storing the voltage value measured at
potPin pin
float voltageValue = 0.; // variable storing the voltage calculated

void setup() {
  Serial.begin(9600);
  pinMode(ledPin, OUTPUT);  // define ledPin pin as an output
}

void loop(){
  potValue = analogRead(potPin); // read and store the read value at
potPin pin

  digitalWrite(ledPin, HIGH);   // turn on the LED
  delay(potValue);              // pause the program during potValue
millisecond
  digitalWrite(ledPin, LOW);    // turn off the LED
  delay(potValue);              // pause the program during potValue
millisecond

  voltageValue = 5. * (potValue / 1023.) ;  // calculate the voltage

  Serial.println(voltageValue); // write the voltage value an a
carriage return
}
```

The code is almost the same as the previous code.

I added a variable to store the calculated voltage. I also added the serial communication stuff, which you see all the time: Serial.begin(9600) to instantiate the serial communication and Serial.println() to write the current calculated voltage value to the serial communication port, followed by a carriage return.

In order to see a result on your computer, you have to turn on the serial monitor, of course. Then, you can read the voltage values.

Calculating the precision

Please note that we are using an ADC here in order to convert an analog value to digital; then, we are making a small calculation on that digital value in order to have a voltage value. This is a very expensive method compared to a basic analog voltage controller.

It means our precision depends on the ADC itself, which has a resolution of 10 bits. It means we can only have 1024 values between 0 V and 5 V. 5 divided by 1024 equals 0.00488, which is approximated.

It basically means we won't be able to distinguish between values such as 2.01 V and 2.01487 V, for instance. However, it should be precise enough for the purposes of our learning.

Again, it was an example because I wanted to point out to you the precision/resolution concept. You have to know and consider it. It will prove very important and could deliver strange results in some cases. At least, you have been warned.

Let's discover another neat way of interacting with the Arduino board.

Introducing Max 6, the graphical programming framework

Now, let me introduce you to the framework known as Max 6. This is a whole universe in itself, but I wanted to write some pages about it in this book because you'll probably come across it in your future projects; maybe you'll be a Max 6 developer one day, like me, or perhaps you'll have to interface your smart physical objects with Max 6-based systems.

The following is one of the patches of my 3D universe project with Max 6:

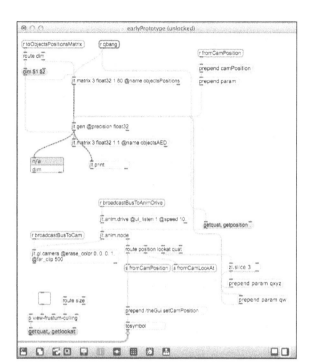

A brief history of Max/MSP

Max is a visual programming language for multimedia purposes. It is actually developed and maintained by Cycling '74. Why call it Max? It was named after Max Matthews (http://en.wikipedia.org/wiki/Max_Mathews), one of the great pioneers of computer music.

The original version of Max was written by Miller Puckette; it was initially an editor named Patcher for Macintosh. He wrote it at **The European Institut de Recherche et Coordination Acoustique/Musique (IRCAM)**, an avant-garde science institute based near the Centre Pompidou in Paris, France.

In 1989, the software was licensed by IRCAM to Opcode Systems, a private company, and ever since then, has been developed and extended by David Zicarelli. In the mid-'90s, Opcode Systems ceased all development for it.

Puckette released a totally free and open source version of Max named Pure Data (often seen as Pd). This version is actually used a lot and maintained by the community that uses it.

Around 1997, a whole module dedicated to sound processing and generation has been added, named **MSP**, for **Max Signal Processing** and, apparently, for the initials of Miller S. Puckette.

Since 1999, the framework commonly known as Max/MSP has been developed and distributed by Cycling '74, Mr. Zicarelli's company.

Because the framework architecture was very flexible, some extensions have progressively been added, such as Jitter (a huge and efficient visual synthesis), Processing, real-time matrix calculations modules, and 3D engine too. This happened around 2003. At that time, Jitter was released and could be acquired separately but required Max, of course.

In 2008, a major update was released under the name Max 5. This version too did not include Jitter natively but as an add-on module.

And the most giant upgrade, in my humble opinion, released in November 2011 as Max 6, included Jitter natively and provided huge improvements such as:

- A redesigned user interface
- A new audio engine compatible with 64-bit OSs
- High-quality sound filter design features
- A new data structure
- New movement handling for 3D models
- New 3D material handling
- The Gen extension

Max 4 was already totally usable and efficient, but I have to give my opinion about Max 6 here. Whatever you have to build, interfaces, complex, or easy communication protocols including HID-based (**HID=human interface device**) USB devices such as Kinect, MIDI, OSC, serial, HTTP, and anything else, 3D-based sound engine or basic standalone applications for Windows or OS X platform, you can make it with Max 6, and it is a safe way to build.

Here is my own short history with Max: I personally began to play with Max 4. I specially built some macro MIDI interfaces for my first hardware MIDI controllers in order to control my software tools in very specific ways. It has taught me much, and it opened my mind to new concepts. I use it all the time, for almost every part of my artistic creation.

Now, let's understand a little bit more about what Max is.

Global concepts

Of course, I hesitated to begin the part about Max 6 in the preceding section. But I guess the little story was a good starting point to describing the framework itself.

What is a graphical programming framework?

A **graphical programming framework** is a programming language that provides a way for users to create programs by manipulating elements graphically instead of by typing text.

Usually, graphical programming languages are also called **visual programming languages**, but I'll use "graphical" because, to many, "visual" is used for the product rendered by frameworks; I mean, the 3D scene for instance. Graphical is more related to **GUI**, that is, **graphical user interface**, which is, from the developer point of view, our editor interface (I mean, the IDE part).

Frameworks using this strong graphical paradigm include many ways of programming in which we can find data, data types, operator and functions, input and output, and a way of connecting hardware too.

Instead of typing long source codes, you add objects and connect them together in order to build software architectures. Think Tinker Toys or Legos.

A global software architecture, which is a system of objects connected and related on our 2D screen, is called **Patch** in the Max world. By the way, other graphical programming frameworks use this term too.

If this paradigm can be understood at first as a way of simplification, it is not the first purpose, I mean that not only is it easier, but it also provides a totally new approach for programmers and non-programmers alike. It also provides a new type of support task. Indeed, if we don't program in the same way we patch, we don't troubleshoot problems in the same way too.

I can quote some other major graphical programming software in our fields:

- **Quartz Composer**: This is a graphical rendering framework for OS X and is available at https://developer.apple.com/technologies/mac/graphics-and-animation.html

- **Reaktor**: This is a DSP and MIDI-processing framework by Native Instruments and is available at http://www.native-instruments.com/#/en/products/producer/reaktor-5

- **Usine**: This is a universal audio software for live and studio recording and is available at http://www.sensomusic.com/usine

- **vvvv**: This is a real-time video synthesis tool for Windows and is available at `http://vvvv.org`

- **SynthMaker**: This is a VST device design for Windows and is available at `http://synthmaker.co.uk`

I'd like to make a special mention of Usine. It is a very interesting and powerful framework that provides graphical programming to design patches usable inside Usine software itself or even as standalone binaries. But one of the particularly powerful features is the fact you can export your patch as a fully-functional and optimized VST plugin. **VST** (**Virtual Studio Technology**) is a powerful standard created by the Steinberg company. It provides a huge list of specifications and is implemented in almost all digital audio workstations. Usine provides a one-click-only export feature that packs your graphically programmed patch into a standard VST plugin for people who haven't even heard about Usine or patching styles. The unique multitouch feature of Usine makes it a very powerful framework too. Then, you can even code your own modules using their C++ **SDKs** (**software development kits**).

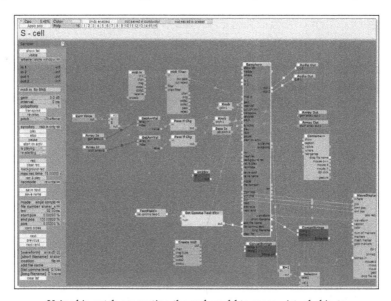

Usine big patch connecting the real world to many virtual objects

Max, for the playground

Max is the playground and the core structure in which everything will be placed, debugged, and shown. This is the place where you put objects, connect them together, create a user interface (UI), and project some visual rendering too.

Here is a screenshot with a very basic patch designed to help you understand where things go:

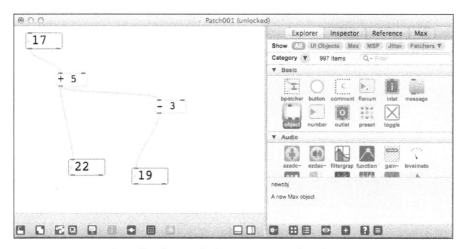

A small and easy calculation system patch with Max 6

As I described, with a graphical programming framework, we don't need to type code to make things happen. Here, I'm just triggering a calculation.

The box with the number **17** inside is a numbox. It holds an integer and it is also a UI object, providing a neat way to change the value by dragging and dropping with a mouse. You then connect the output of one object to the input of another. Now when you change the value, it is sent through the wire to the object connected to the numboxes. Magic!

You see two other objects. One with a:

*   + sign inside followed by the number **5**
*   - sign inside followed by the number **3**

Each one takes the number sent to them and makes the calculation of + 5 and - 3 respectively.

You can see two other numboxes displaying basically the resulting numbers sent by the objects with the **+** and **–** signs.

Are you still with me? I guess so. Max 6 provides a very well documented help system with all references to each object and is directly available in the playground. It is good to tell that to students when you teach them this framework, because it really helps the students teach themselves. Indeed, they can be almost autonomous in seeking answers to small questions and about stuff they have forgotten but don't dare to ask.

Max part provides quite an advanced task scheduler, and some objects can even modify priority to, say, `defer` and `deferlow` for a neat granularity of priorities inside your patch, for instance, for the UI aspect and the calculation core aspect that each require very different scheduling.

Max gives us a nifty debugging system too with a console-like window called the **Max window**.

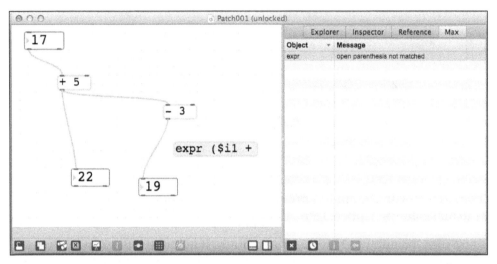

The Max window showing debugging information about the expr object's error in the patch

Max drives many things. Indeed, it is Max that owns and leads the access to all modules, activated or not, provides autocompletion when you create new objects, and also gives access to many things that can extend the power of Max, such as:

- JavaScript API to Max itself and specific parts, such as Jitter, too
- Java through the mxj object that instantiates directly inside Max 6 Java classes
- MSP core engine for everything related to signal rate stuff, including audio
- Jitter core engine for everything related to matrix processing and much more, such as visuals and video
- Gen engine for efficient and on-the-fly code compilation directly from the patch

This is not an exhaustive list, but it gives you an insight of what Max provides.

Let's check the other modules.

MSP, for sound

Where Max objects communicate by sending messages triggered by user or by the scheduler itself, MSP is the core engine that calculates signals at any particular instant, as written in the documentation.

Even if we can patch (or connect) MSP objects in the same way as pure Max objects, the concept underneath is different. At each moment, a signal element is calculated, making an almost continuous data flow through what we call a signal network. The signal network is easy to identify in the patcher window; the wires are different.

Here is an image of a very simple patch producing a cosine-based audio wave in your ears:

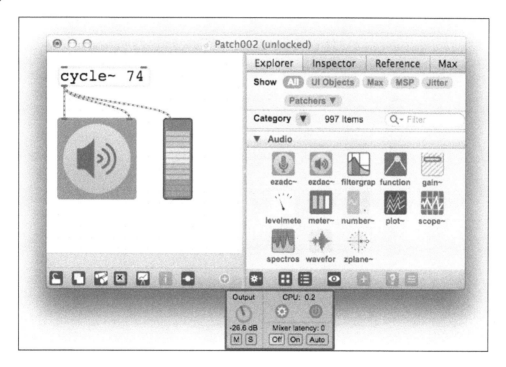

Indeed, even the patch cords have a different look, showing cool, striped yellow-and-black, bee-like colors, and the names of the MSP objects contain a tilde ~ as a suffix, symbolizing…a wave of course!

The signal rate is driven by the audio sampling rate and some dark parameters in the MSP core settings window. I won't describe that, but you have to know that Max usually provides, by default, parameters related to your soundcard, which include the sampling rate (44110 Hz, the standard sampling rate for audio CDs, means a fast processing rate of 44100 times per second for each audio channel).

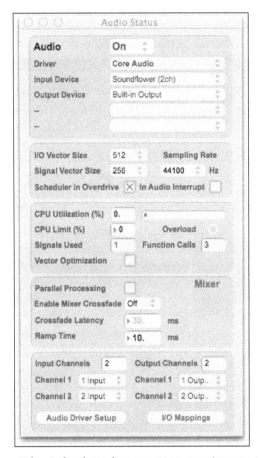

The Audio Status window is the place where you set up some important MSP parameters

Jitter, for visuals

Jitter is the core engine for everything related to visual processing and synthesis in Max 6.

It provides a very efficient framework of matrix processing initially designed for fast pixel value calculations to display pictures, animated or not.

We are talking about matrix calculation for everything related with Jitter processing matrices. And indeed, if you need to trigger very fast calculations of huge arrays in Max 6, you can use Jitter for that, even if you don't need to display any visuals.

Jitter provides much more than only matrix calculation. It gives full access to an OpenGL (http://en.wikipedia.org/wiki/OpenGL) implementation that works at the speed of the light. It also provides a way for designing and handling particle systems, 3D worlds, OpenGL materials, and physics-based animation. Pixel processing is also one of the powerful features provided with many objects designed and optimized for pixel processing itself.

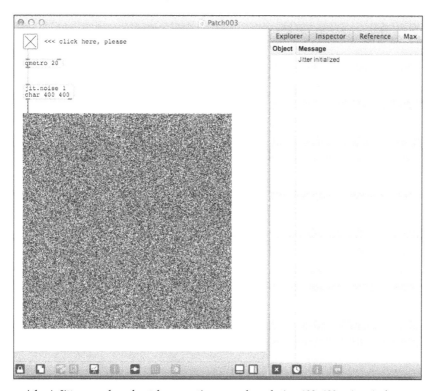

A basic Jitter-core-based patch generating a good resolution 400x400 noise pixel map

In order to summarize this massive load of information, Max schedules events or waits for the user to trigger something, MSP (for audio signal processing) — as soon as it is activated — calculates signal elements at each instant in its signal networks, and Jitter processes calculations when Jitter objects are triggered by **bangs**.

Indeed, Jitter objects need to be triggered in order to do their jobs, which can be very different, such as popping out a matrix that contains pixel color values, matrix processing for each cell of a matrix, and popping out the resulting matrix, for instance.

Bangs are special messages used to kinda say *"Hey, let's start your job!"* to objects. Objects in Max can behave differently, but almost every one can understand the bang message.

In **Patch003** (pictured in the previous screenshot), the Max object qmetro provides a bang every 20 ms from a low priority scheduler queue to a Jitter object named jit.noise. This latter object calculates a matrix filled with a random value in each cell. Then, the result goes through a new green-and-black-striped patch cord to a UI object in which we can see a name, the jit.pwindow, a kind of display we can include in our patchers.

Jitter can be controlled via powerful Java and JavaScript APIs for some tasks that require typing big loops in code , which are easy to design using code.

Still here?

For the bravest among the brave, some other rows about Gen, the latest and most efficient module of Max 6.

Gen, for a new approach to code generation

If you understood that there was a kind of compilation/execution behind our patches, I'd disappoint you by saying it doesn't really work like that. Even if everything works real time, there isn't a real compilation.

By the way, there are many ways to design patch bits using code, with JavaScript for instance. Directly inside Max patcher, you can create a .js object and put your JavaScript code inside; it is indeed compiled on the fly (it is called **JS JIT** compiler, for JavaScript just-in-time compiler). It is really fast. Believe me, I tested it a lot and compared it to many other frameworks. So, as the documentation said, "we are not confined to writing Max externals in C" even if it is totally possible using the Max 6 SDK (http://cycling74.com/products/sdk).

Gen is a totally new concept.

Gen provides a way of patching patch bits that are compiled on the fly, and this is a real compilation from your patch. It provides a new type of patcher with specific objects, quite similar to Max objects.

It works for MSP, with the gen~ Max object, providing a neat way to design signal-rate related to audio patches architecture. You can design DSP and sound generators like that. The gen~ patches are like a zoom in time; you have to consider them as sample processors. Each sample is processed by those patches inside the gen~ patchers. There are smart objects to accumulate things over time, of course, in order to have signal processing windows of time.

It works also for Jitter with three main Max objects:

- `jit.gen` is the fast matrix processor, processing each cell of a matrix at each turn

- `jit.pix` is the CPU-based pixel processor, processing each pixel of a pixel map

- `jit.gl.pix` is the GPU-based version of `jit.pix`

A GPU (graphics processor unit), and is basically a dedicated graphics processor on your video card. Usually, and this is a whole different universe, OpenGL pipeline provides an easy way to modify pixels from the software definitions to the screen just before they are displayed on the screen. It is called **shader process**.

You may already know that term in relation with the world of gaming. These are those shaders that are some of the last steps to improving graphics and visual renders in our games too.

Shaders are basically small programs that can be modified on the fly by passing arguments processed by the GPU itself. These small programs use specific languages and run vary fast on dedicated processors on our graphic cards.

Max 6 + Gen provides direct access to this part of the pipeline by patching only; if we don't want to write shaders based on **OpenGL GLSL** (http://www.opengl.org/documentation/glsl), **Microsoft DirectX HLSL** (http://msdn.microsoft.com/en-us/library/bb509635(v=VS.85).aspx), or **Nvidia Cg** (http://http.developer.nvidia.com/CgTutorial/cg_tutorial_chapter01.html), Gen is your friend.

All patches based on `jit.gl.pix` are specifically compiled and sent for GPU-based execution.

You can then design your own fragment shaders (or pixel shaders) by patching and you can even grab the source code in GLSL or WebGL language in order to use it in another framework, for instance.

Geometry shaders aren't available using Gen, but with other Jitter objects they already exists.

I guess I lost some of you. Relax I won't ask you questions about Gen in Arduino exams!

Summarizing everything in one table

Everything related to Max 6 is on the Cycling 74's website at `http://cycling74.com`. Also, almost 99 percent of the documentation is online too, at `http://cycling74.com/docs/max6/dynamic/c74_docs.html#docintro`.

The following table summarizes everything we did until now:

Parts	What?	Cable color	Distinctive sign
Max	The playground	Gray by default and no stripes	Basic names
MSP	Everything related to audio and signal rate	Yellow-and-black stripes	~ suffixed to the name
			signal-rate processing
Jitter	Everything related to visuals and matrices	Green-and-black stripes for matrix cables	`jit.` prefixed to the name
		Blue-and-black stripes for pixel map cables	
Gen	Specific patchers (DSP-related and matrix and texture processing) compiled on the fly	Like MSP for `gen~` and Jitter for `jit.gen`, `jit.pix`, and `jit.gl.pix`	Very very fast!

Installing Max 6

Max 6 is available as a 30-day trial. Installing Max 6 is quite easy as it comes with an installer for both platforms, Windows and OS X, downloadable at `http://cycling74.com/downloads`. Download and install it. Then, launch it. That's all. (The following examples will only work when you have installed Max.)

You should see a blank playground

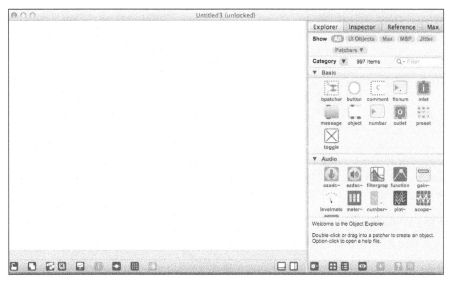

Max 6 blank-page anxiety can occur right now, can't it?

The very first patch

Here is a basic patch you can find also in the `Chapter06/` folder under the name `Patcher004_Arduino.maxpat`. Usually, if you double-click on it, it is opened directly by Max 6.

This patch is a very basic one, but not that basic actually!

It is a basic noise-based sequencer modifying an oscillator's frequency regularly in real time. This produces a sequence of strange sounds, more or less pretty, the modifications of the frequency being controlled by chance. So, turn on your speakers and the patch will produce sounds.

The noise-based sequencer

Basically, patches are stored in files. You can share patches with other friends quite easily. Of course, bigger projects would involve some dependency issues; if you added some libraries to your Max 6 framework, if you use them in a patch, or if you basically send your patch files to a friend who doesn't have those libraries installed, your friend will have some errors in the Max Window. I won't describe these kinds of issues here, but I wanted to warn you.

Other neat ways to share patches in the Max 6 world are the copy/paste and copy compressed features. Indeed, if you select objects in your patcher (whatever the layer, including a subpatcher, inside a subpatcher, and so on) and go to **Edit** | **Copy**, text-based content is put in your clipboard. This can then be repasted into another patcher or inside a text file.

The smartest way is the use of copy compress, which as the well-chosen name means, copies and compresses the JSON code to something much more compact and easy to copy into the text area on forums, for instance.

Wait, let me show you what it looks like.

I just selected all objects in my patch and went to **Edit** | **Copy Compressed**.

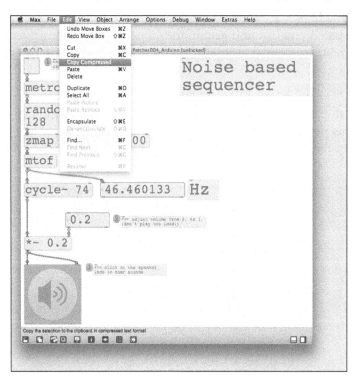

The copy compressed feature

And the following figure is the result of pasting directly into a text file.

Those familiar with HTML would notice something funny; Cycling '74 developers include two HTML tags (pre and code) in order to directly provide code that is pastable inside a text field on (any) forums on the Web.

```
<pre><code>
----------begin_max5_patcher----------
989.3oc2XtrbihCEFds8SgF1LWJOT5Jf5J6lMyp4EXpo5RFTvjFjbChLIoqN
O6ijv33jFmPH1zclMP4CxGN5S+mK1eY4hf05ajMAf0.9avhEeY4hEdSNCK18
4EAUhaRKEM9kEjpqpjJSvptmYj2X71+KcQiDrVzHy.Mx02JUox59UUVnjo5V
keo3cFKx7eQ85q9cJsekWpUFknR5ezenaqKdvKaElzMEp70VKSMcwLkBCgq.
b2EBC6tgfnPH3eNveME248GMNjfwIv39mUmu1Y24hNKp1pBUoz32pnGLpaM8
VgNiec4R2kUuQl8m2EL.KHSkEQdVPHnPFBhwjUfDpyBi9dAHWbwEfzxhz0Az
JfYiDzrUJ9jrFTpy0.iFrQJpAMVkTVyn0VnIxSD2KqXrH2MLj3oK7HrDg2Cs
YCiUxlFQt7avHYHYEdLXnKfL2tU1wfff8a1gHTBriP3vDTTDyl74QFFeLHEG
xsxMLc2yVmmpK006716qBCYDJKJdkmdfCSkeDDwCBQzIVKJxtpsw.tVW1VIA
WVqqrgkSGhBA+RlV8yFv1RwsVKZqDsM6m90wpJI7IpJwwLGXn3j8Y4XB6ciz
DOjzDd5kl3trVJlYKwEwI++SaltQZqSVXbxwRQqJciuhYsPkYko1mJ195aDS
RlnrLwKAsJMTBAmvc0KguaDknADkjnSunLJZ+vJuCEiWVpsNYHTQmDpr9SXk
1ff0BU9y2mA668RQOTyq2zgxqshZ661Hq+nTIVWSe0vAfKgDlDEwi6gqkVEU
1c1ijf1sduQzqQWhm.YUx+0xwuQV9a2aimgJXh4SB3ME4JQ4yRZupjlfBsPl
gPVNS5F6geDgJw1SGxiQnuSZOb7YV6gw6lp9v9smNs2YRQIuKSjd+P7hD7Bm
9612Ew36u8Zpr.Oc5+zaSKk2ChoCsOPm0T.h8vceJ.G+iPJvQfTkQe4.7AEe
5ae00xJhrmL72PsAz4FL2UYG.BZaTDamEAffvgnD8bQIjW+P4LmVJY5Xhctw
zt4EQ3jwNiHaRLqv4uWBa1CqPdLOBQ2W.JF+8MsynyyKkCM0yaaq10ok1ecT
MRl2DHaTTqOVhyjz.uXyVOTh3NM.iy06kd89zq2exeApOzb1eL0Zr6uz9MSe
UVvCgVlrwTnDlBs5fE4ZUcvh1TjkIUGdrlUz3NoyNd6ywFOtIwe43wN8H.MK
wyS15OS7LO74IupiDOje.im447Zb5434SOOhvY9xtFiXlMeQyXx0Qy2YEaLg
CchgSWkZw1sWKqa14SejX6jcU2+KRzpkc+V9tO58XPs75h90Guz4sut7+rKF
1bK
----------end_max5_patcher----------
</code></pre>
```

Copy-compressed code

So you can also copy that code into your clipboard and paste it into a new patch. You create a new empty patch by by going to **File** | **New** (or hitting *Ctrl + N* on Windows and *command + N* on OS X).

Playing sounds with the patch

As you can see, I put some comments in the patcher. You can follow them in order to produce some electronic sounds from your computer.

Before you begin, be sure to lock the patch by clicking on the padlock icon in the lower-left corner. To hear the results of the patch, you'll also need to click on the speaker icon. To zoom out, go to the **View** menu and click on **Zoom Out**.

First, note and check the toggle at the top. It will send the value 1 to the connected object metro.

A metro is a pure Max object that sends a bang every *n* milliseconds. Here, I hardcoded an argument: 100. As soon as the metro receives the message 1 from the toggle, it begins to be active and, following the Max timing scheduler, it will send its bangs every 100 ms to the next connected object.

When the random object receives a bang, it pops out a random integer from within a range. Here, I put 128, which means random will send values from 0 to 127. Directly after random, I put a zmap object that works like a scaler. I harcoded four arguments, minimum and maximum values for inputs and minimum and maximum values for output.

Basically, here, zmap maps my values 0 to 127 sent by random to another values from 20 to 100. It produces an implicit stretch and loss of resolution that I like.

Then, this resulting number is sent to the famous and important mtof object. This converts a MIDI note pitch standard to a frequency according to the MIDI standard. It is often used to go from the MIDI world into the real sound world. You can also read the frequency in the UI object flonum displaying the frequency as a float number in Hz (hertz, a measure of frequency).

Then, at last, this frequency is sent to the cycle~ object, producing a signal (check the yellow-and-black striped cord). Sending numbers to this object makes it to change the frequency of the signal produced. This one is multiplied by a signal multiply operator *~, producing another signal but with a lower amplitude to protect our precious ears.

The last destination of that signal is the big gray box on which you have to click once in order to hear or not hear the sounds produced by the upper signal network.

Now you're ready to check the toggle box. Activate the speaker icon by clicking on the gray box, and then you can dance. Actually, electronic sounds produced are a bit shuffly about the frequency (that is, the note) but it can be interesting.

Of course, controlling this cheap patch with the Arduino in order to not use the mouse/cursor would be very great.

Let's do that with the same circuit that we designed previously.

Controlling software using hardware

Coming from pure digital realms where everything can be wrapped into software and virtual worlds, we often need physical interfaces. This can sound like a paradox; we want everything in one place, but that place is so small and user-unfriendly for everything related to pure creation and feelings that we need more or less big external (physical) interfaces. I love this paradox.

But, why do we need such interfaces? Sometimes, the old mouse and QWERTY keyboard don't cut it. Our computers are fast, but these interfaces to control our programs are slow and clunky.

We need interfaces between the real world and the virtual world. Whatever they are, we need them to focus on our final purpose, which is usually not the interface or even the software itself.

Personally, I write books and teach art-related technology courses, but as a live performer, I need to focus on the final rendering. While performing, I want to black-box as much as possible the technology under the hood. I want to feel more than I want to calculate. I want a controller interface to help me operate at the speed and level of flexibility to make the types of changes I want.

As I already said in this book, I needed a huge MIDI controller, heavy, solid, and complex, in order to control only one software on my computer. So, I built Protodeck (`http://julienbayle.net/protodeck`). This was my interface.

So, how can we use Arduino to control software? I guess you have just a part of your answer because we already sent data to our computer by turning a potentiometer.

Let's improve our Max 6 patch to make it receive our Arduino's data while we turn the potentiometer.

Improving the sequencer and connecting the Arduino

We are going to create a very cheap and basic project that will involve our Arduino board as a small sound controller. Indeed, we'll directly use the firmware we just designed with the potentiometer, and then we'll modify our patch. This is a very useful base for you to continue to build things and even create bigger controller machines.

Let's connect the Arduino to Max 6

Arduino can communicate using the serial protocol. We already did that. Our latest firmware already does that, sending the voltage value.

Let's modify it a bit and make it send only the analog value read, within the range 0 to 1023. Here is the code, available in Chapter06/maxController:

```
int potPin = 0;      // pin number where the potentiometer is connected
int potValue = 0 ;   // variable storing the voltage value measured at
potPin pin

void setup() {
  Serial.begin(9600);
}

void loop(){
  potValue = analogRead(potPin); // read and store the read value at
potPin pin
  Serial.println(potValue); // write the voltage value an a carriage
return

  delay(2);      // this small break waits for the ADC to stabilize is
often used
}
```

I removed everything unnecessary and added a delay of 2 ms at the end of the loop (before the loop restarts) This is often used with analog input and especially ADC. It provides a break to let it stabilize a bit. I didn't do that in previous code involving analog read because there were already two delay() methods involved in the LED blinking.

This basic one sends the value read at the analog input pin where the potentiometer is connected. No more, but no less.

Now, let's learn how to receive that somewhere other than the Serial Monitor of our precious IDE, especially in Max 6.

The serial object in Max 6

There is a Max object named serial. It provides a way to communicate using a serial port with any other type of device using serial communication.

The next figure describes the new Max 6 patch including the part necessary to communicate with our small hardware controller.

Now, let's plug the Arduino in, if this has not been done already, and upload the maxController firmware.

Be careful to switch off serial monitoring for the IDE. Otherwise, there would be a conflict on your computer; only one serial communication can be instantiated on one port.

Then here is another patch you can find, also in the `Chapter06/` folder, with the name `Patcher005_Arduino.maxpat`.

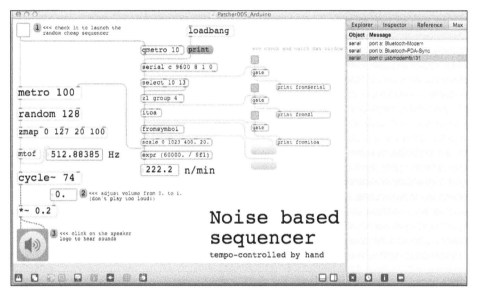

The Max patch including the Arduino communication module

Double-click on the file, and you'll see this patch.

Let's describe it a bit. I added everything in green and orange.

Everything necessary to understand the Arduino messages and to convert them in terms understandable easily by our sequencer patch is in green. Some very useful helpers that are able to write to the Max window at every step of the data flow, from raw to converted data, are in orange.

Let's describe both parts, beginning with the helpers.

Tracing and Debugging easily in Max 6

Max 6 provides many ways to debug and trace things. I won't describe them all in this Arduino book, but some need a few words.

Check your patch, especially the orange-colored objects.

print objects are the way to send messages directly to the Max window. Everything sent to them is written to the Max window as soon it has been received. The argument you can pass to these objects is very useful too; it helps to discern which print object sends what in cases where you use more than one print object. This is the case here and check: I name all my print objects considering the object from which comes the message:

- fromSerial: This is for all messages coming from the serial object itself
- fromZl: This is for all messages coming from the zl object
- fromitoa: This is for all messages coming from the itoa object
- fromLastStep: This is for all messages coming from the fromsymbol object

The gate objects are just small doors, gates that we can enable or disable by sending 1 or 0 to the leftmost input. The toggle objects are nice UI objects to do that by clicking. As soon as you check the toggle, the related gate object will let any message sent to the right input pass through them to the only one output.

We are going to use this trace system in several minutes.

Understanding Arduino messages in Max 6

What is required to be understood is that the previous toggle is now connected to a new qmetro object too. This is the low priority metro equivalent. Indeed, this one will poll the serial object every 20 ms, and considering how our Arduino's firmware currently works by sending the analog value read at every turn in the loop, even if this polling lags a bit, it won't matter; the next turn, the update will occur.

The serial object is the important one here.

I hardcoded some parameters related to serial communication with the Arduino:

- 9600 sets the clock to 9600 bauds
- 8 sets the word length at 8 bit
- 1 means there is a stop bit
- 0 means there is no parity (parity is sometimes useful in error checking)

This object needs to be banged in order to provide the current content of the serial port buffer. This is the reason why I feed it by the `qmetro` object.

The `serial` object pops out a raw list of values. Those values need to be a bit parsed and organized before reading the analog value sent. This is what the `select`, `zl`, `itoa`, and `fromsymbol` objects stand for.

 Directly read the help information for any object in Max 6 by pushing the *Alt* key on your keyboard and then clicking on the object.

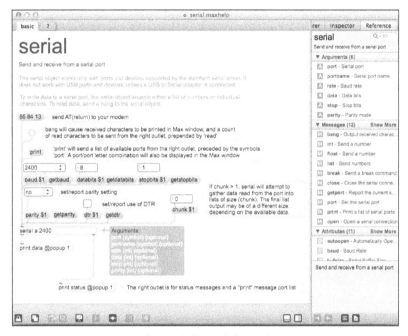

The serial object's help patch

Every 20 ms, if the serial communication has been instantiated, the `serial` object will provide what will be sent by the Arduino, the current and most recently read analog value of the pin where the potentiometer is connected. This value going from 0 to 1023, I'm using a `scale` object as I did with the `zmap` object for the sequencer/sound part of the patch. This `scale` object recasts the scale of values from 0 to 1023 at input to an inverted range of 300 down to 20, letting the range to go opposite direction (be careful, current and future Max patchers, `zmap` doesn't behave like that). I did that in order to define the maximum range of the note-per-minute rate. The `expr` object calculates this. `qmetro` needs the interval between two bangs. I'm making this vary between 400 ms and 20 ms while turning my potentiometer. Then, I calculate the note-per-minute rate and display it in another `flonum` UI object.

Then, I also added this strange `loadbang` object and the `print` one. `loadbang` is the specific object that sends a bang as soon as the patcher is opened by Max 6. It is often used to initialize some variable inside our patcher, a bit like we are doing with the declarations in the first rows of our Arduino sketches.

`print` is only text inside an object named `message`. Usually, each Max 6 object can understand specific messages. You can create a new empty message by typing `m` anywhere in a patcher. Then, with the autocomplete feature, you can fill it with text by selecting it and clicking on it again.

Here, as soon as the patch is loaded and begins to run, the `serial` object receives the print message triggered by `loadbang`. The `serial` object is able to send the list of all serial port messages to the computer that runs the patch to the console (that is, the Max window). This happens when we send the print message to it. Check the Max window of the figure showing the `Patcher005_Arduino.maxpat` patch.

We can see a list of…things. `serial` pops out a list of serial port letter abbreviations with the corresponding serial ports often representing the hardware name. Here, as we already saw in the Arduino IDE, the one corresponding to the Arduino is `usbmodemfa131`.

The corresponding reference in Max is the letter `c` on my computer. This is only an internal reference.

Explorer	Inspector	Reference	Max
Object	**Message**		
serial	port a: Bluetooth-Modem		
serial	port b: Bluetooth-PDA-Sync		
serial	port c: usbmodemfa131		

Result of the print message sent to the serial object: the list of port letters / names of serial ports

Let's change the hardcoded letter put as argument for the `serial` object in the patch.

Select the `serial` object. Then, re-click inside and swap `a` with the letter corresponding to the Arduino serial port on your computer. As soon as you hit *Enter*, the object is instantiated again with new parameters.

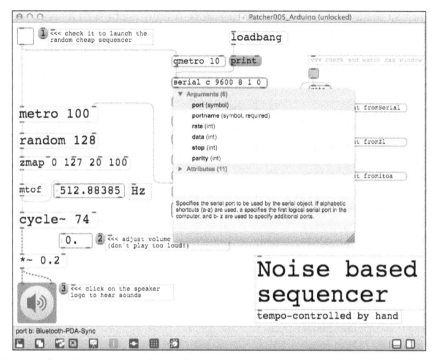

Changing the reference letter in the serial object to match the one corresponding to the serial port of the Arduino

Now, everything is ready. Check the toggle, enable the gray box with the speaker, and turn your potentiometer. You are going to hear your strange noises from the sequencer, and you can now change the note rate (I mean the interval between each sound) because I abusively used the term note to fit better to the sequencer's usual definition.

What is really sent on the wire?

You will have noticed that, as usual, I mentioned the series of objects: `select`, `zl`, `itoa`, and `fromsymbol`. The time has come to explain them.

When you use the `Serial.println()` function in your Arduino's firmware source code, the Arduino doesn't send only the value passed as argument to the function. Check the first orange toggle at the top of the series of toggle/gate systems.

Explorer	Inspector	Reference	Max
Object	**Message**		
fromSerial	13		
fromSerial	10		
fromSerial	51		
fromSerial	53		
fromSerial	48		
fromSerial	13		
fromSerial	10		
fromSerial	51		
fromSerial	53		
fromSerial	48		
fromSerial	13		
fromSerial	10		
fromSerial	51		
fromSerial	53		

The serial object pops out strange series of numbers

You can see the name of the printing object in the first column named **Object**, and in the **Message** column, the message sent by the related object. And we can see the `serial` object popping out strange series of numbers in a repetitive way: **51, 53, 48, 13, 10**, and so on.

 Arduino transmits its values as ASCII, exactly as if we were typing them on our computer.

This is very important. Let's check the *Appendix E, ASCII Table*, in order to find the corresponding characters:

- 51 means the character 3
- 53 means 5
- 48 means 0
- 13 means a carriage return
- 10 means line feed, which itself means new line

Of course, I cheated a bit by sorting the series as I did. I knew about the `10 13` couple of numbers. It is a usual marker meaning *a carriage return followed by a new line.*

So it seems that my Arduino sent a message a bit like this:

```
350<CR><LF>
```

Here, `<CR>` and `<LF>` are carriage return and new line characters.

If I had used the `Serial.print()` function instead of `Serial.println()`, I wouldn't have had the same result. Indeed, the `Serial.print()` version doesn't add the `<CR>` and `<NL>` characters at the end of a message. How could I have known whether 3, 5, or 0 would be the first character if I didn't have an end marker?

The design pattern to keep in mind is as follows:

- Build the message
- Send the message after it is completely built (using the `Serial.println()` function.)

If you want to send it while building it, here's what you can use:

- Send the first byte using `Serial.print()`
- Send the second byte using `Serial.print()`
- Continue to send until the end
- Send the `<CR><LF>` at the end by using `Serial.println()` with no argument

Extracting only the payload?

In many fields related to communication, we talk about payload. This is the message, the purpose of the communication itself. Everything else is very important but can be understood as a carrier; without these signals and semaphores, the message couldn't travel. However, we are interested in the message itself.

We need to parse the message coming from the serial object.

We have to accumulate each ASCII code into the same message, and when we detect the `<CR><LF>` sequence, we have to pop out the message block and then restart the process.

This is done with the `select` and `zl` objects.

`select` is able to detect messages equaling one of its arguments. When `select 10 13` receives a 10, it will send a bang to the first output. If it is a 13, it will send a bang to the second output. Then, if anything else comes, it will just pass the message from the last output to the right.

z1 is such a powerful list processor with so many usage scenarios that it would make up a book by itself! Using argument operator, we can even use it to parse the data, cut lists into pieces, and much more. Here, with the group 4 argument, z1 receives an initial message and stores it; when it receives a second message, it stores the message, and so on, until the fourth message. At the precise moment that this is received, it will send a bigger message composed of the four messages received and stored. Then, it clears its memory.

Here, if we check the corresponding toggle and watch the Max window, we can see **51 53 48** repeated several times and sent by the z1 object.

The z1 object does a great job; it passes all ASCII characters except <CR> and <LF>, and as soon as it receives <LF>, z1 sends a bang. We have just built a message processor that *resets* the z1 buffer each time it receives <LF>, that is, when a new message is going to be sent.

Explorer	Inspector	Reference	Max
Object	**Message**		
fromZl	51 53 48		
fromZl	51 53 48		
fromZl	51 53 48		
fromZl	51 53 48		
fromZl	51 53 48		
fromZl	51 53 48		
fromZl	51 53 48		
fromZl	51 53 48		
fromZl	51 53 48		
fromZl	51 53 48		
fromZl	51 53 48		
fromZl	51 53 48		
fromZl	51 53 48		
fromZl	51 53 48		
fromZl	51 53 48		

The zl list processor pops out a series of integers

ASCII conversions and symbols

We have now a series of three integers directly equaling the ASCII message sent by the Arduino, in my case, 51 53 48.

If you turn the potentiometer, you'll change this series, of course.

But look at this, where is the value between 0 and 1023 we so expected? We have to convert the ASCII integer message into a real character one. This can be done using the `itoa` object (which means integer to ASCII).

Check the related toggle, and watch the Max window.

Explorer	Inspector	Reference	Max
Object	**Message**		
fromitoa	350		
fromitoa	350		
fromitoa	350		
fromitoa	350		
fromitoa	350		
fromitoa	350		
fromitoa	350		
fromitoa	350		
fromitoa	350		
fromitoa	350		
fromitoa	350		

Here is our important value

This value is the important one; it is the message sent by the Arduino over the wire and is transmitted as a symbol. You cannot distinguish a symbol from another type of message, such as an integer or a float in the Max window.

I placed two empty messages in the patch. Those are really useful for debugging purposes too. I connect them to the `itoa` and `fromsymbol` objects to their right input. Each time you send a message to another message on its right input, the value of the destination message is changed by the content of the other one. We can then display what message is really sent by `itoa` and `fromsymbol`.

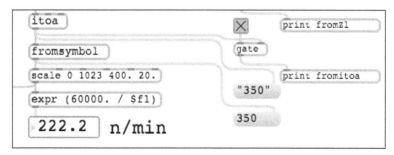

"350" doesn't equal exactly 350

`fromsymbol` transforms each symbol into its component parts, which here make up an integer, `350`.

This final value is the one we can use with every object able to understand and process numbers. This value is scaled by the scale object and sent, at last, to the metro object. Turning the potentiometer changes the value sent, and depending upon the value, the metro sends bangs faster or slower.

This long example taught you two main things:

- You have to carefully know what is sent and received
- How an Arduino communicates

Now, let's move on to some other examples relating to analog inputs.

Playing with sensors

What I don't want to write in this book is a big catalog. Instead of that, I want to give you keys and the feel of all the concepts. Of course, we have to be precise and learn about particular techniques you didn't invent yourself, but I especially want you to learn best practices, to think about huge projects by yourself, and to be able to have a global vision.

I'll give you some examples here, but I won't cover every type of sensor for the previously mentioned reason.

Measuring distances

When I design installations for others or myself, I often have the idea of measuring distance between moving things and a fixed point. Imagine you want to create a system with a variable light intensity depending on the proximity of some visitors.

I used to play with a Sharp GP2Y0A02YK infrared long range sensor.

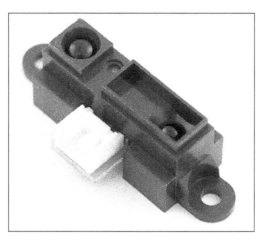

The infrared Sharp GP2Y0A-family sensor

This cool analog sensor provides good results for distances from 20 to 150 cm. There are other types of sensors on the market, but I like this one for its stability.

As with any distance sensors, the subject/target has to theoretically be perpendicular to the infrared beam's direction for maximum accuracy, but in the real world, it works fine even otherwise.

The datasheet is a first object to take care about.

Reading a datasheet?

First, you have to find the datasheet. A search engine can help a lot. This sensor's datasheet is at `http://sharp-world.com/products/devvice/lineup/data/pdf/datasheet/gp2y0a02_e.pdf`.

You don't have to understand everything. I know some fellows would blame me here for not explaining the datasheet, but I want my students to be relaxed about that. You have to filter information.

Ready? Let's go!

Generally, on the first page, you have all the features summarized.

Here, we can see this sensor seems to be quite independent considering the color of the target. Ok, good. The distance output type is very important here. Indeed, it means it outputs the distance directly and needs no additional circuitry to utilize its analog data output.

There are often some schematics of all dimensions of the outline of the sensor. This can be very useful if you want to be sure the sensor fits your box or installation before ordering it.

In the next figure, we can see a graph. This is a curve illustrating how the output voltage varies according to the distance of the target.

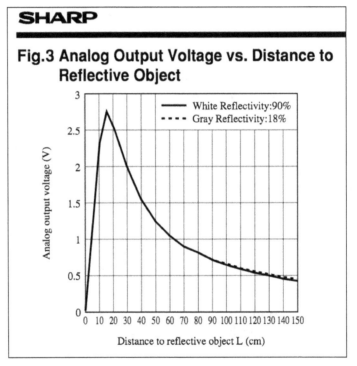

Mathematical relation between distance and analog output voltage from the sensor

This information is precious. Indeed, as we discussed in the previous chapter, a sensor converts a physical parameter into something measurable by Arduino (or any other type of equipment). Here, a distance is converted into a voltage.

Because we measure the voltage with the analog input of our Arduino board, we need to know how the conversion works. And I'm going to use a shortcut here because I made the calculation for you.

Basically, I used another graph similar to the one we saw but mathematically generated. We need a formula to code our firmware.

If the output voltage increases, the distance decreases following *a kind of* exponential function. I had been in touch with some Sharp engineers at some point and they confirmed my thoughts about the type of formula, providing me with this:

$$D = \frac{(a + bV)}{(1 + cV + dV^2)}$$

Here, D is the distance in centimeters and V the voltage measured; and a = 0.008271, b = 939.65, c = -3.398, and d = 17.339

This formula will be included in Arduino's logic in order to make it directly provide the distance to anyone who would like to know it. We could also make this calculation on the other side of the communication chain, in a Max 6 patch for instance, or even in Processing. Either way, you want to make sure your distance parameter data scales well when comparing the output from the sensor to the input where that data will be used.

Let's wire things

This next circuit will remind you very much of the previous one. Indeed, the range sensor replaces the potentiometer, but it is wired in exactly the same way:

- The Vcc and ground of the Arduino board connected respectively to +5 V and ground
- The signal legs connected to the analog input 0

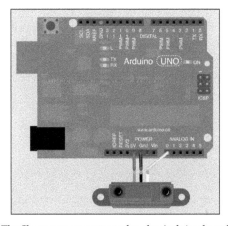

The Sharp sensor connected to the Arduino board

The circuit diagram is as follows:

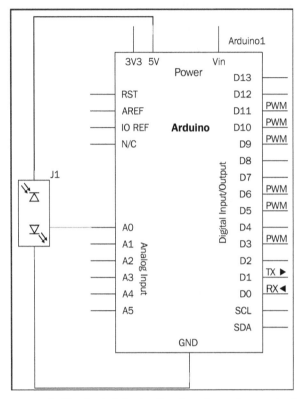

The sensor range supplied by the Arduino itself and sending voltage to the Analog Input 0

Coding the firmware

The following code is the firmware I designed:

```
int sensorPin = 0; // pin number where the SHARP GP2Y0A02YK is
connected
int sensorValue = 0
int distanceCalculated = 0;   // variable storing the distance
calculated
int v = 0;              // variable storing the calculated voltage
```

```
// our formula's constants
const int a = 0.008271;
const int b = 939.65;
const int c = -3.398;
const int d = 17.339;


void setup() {
  Serial.begin(9600);
}

void loop(){
  sensorValue = analogRead(sensorPin);
  v = 5. * (sensorValue / 1023.) ;   // calculate the voltage
  distanceCalculated = ((a + b * v) / (1. + c * v + d * v * v) );

  Serial.println(distanceCalculated);

  delay(2);
}
```

Is it not gratifying to know you understood every line of this code? Just in case though, I will provide a brief explanation.

I need some variables to store the sensor value (that is, the values from 0 to 1023) coming from the ADC. Then, I need to store the voltage calculated from the sensor value, and of course, the distance calculated from the voltage value.

I only initiate serial communication in the setup() function. Then, I make every calculation in the loop() method.

I started by reading the current ADC value measured and encoded from the sensor pin. I use this value to calculate the voltage using the formula we already used in a previous firmware. Then, I inject this voltage value into the formula for the Sharp sensor and I have the distance.

At last, I send the distance calculated through serial communication with the Serial.println() function.

Reading the distance in Max 6

`Patcher006_Arduino.maxpat` is the patch related to this distance measurement project. Here it is:

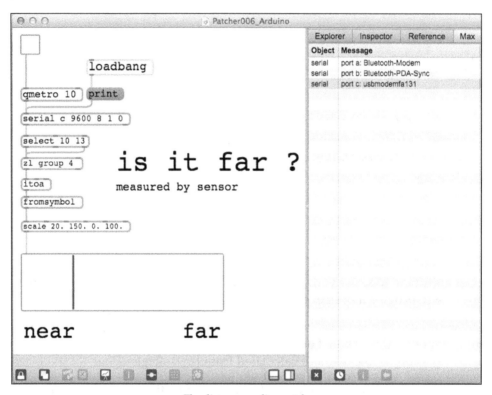

The distance reading patch

As we learnt previously, this patcher contains the whole design pattern to read messages coming from the Arduino board.

The only new thing here is the strange UI element at the bottom. It is called a **slider**. Usually, sliders are used to control things. Indeed, when you click and drag a slider object, it pops out values. It looks like sliders on mixing boards or lighting dimmers, which provide control over some parameters.

Obviously, because I want to transmit a lot of data myself here, I'm using this slider object as a display device and not as a control device. Indeed, the slider object also owns an input port. If you send a number to a slider, the slider takes it and updates its internal current value; it also transmits the value received. I'm only using it here as a display.

Each object in Max 6 has its own parameters. Of course a lot of parameters are common to all objects, but some aren't. In order to check those parameters:

- Select the object
- Check the inspector by choosing the **Inspector** tab or typing *Ctrl + I* on Windows or *command + I* on OS X

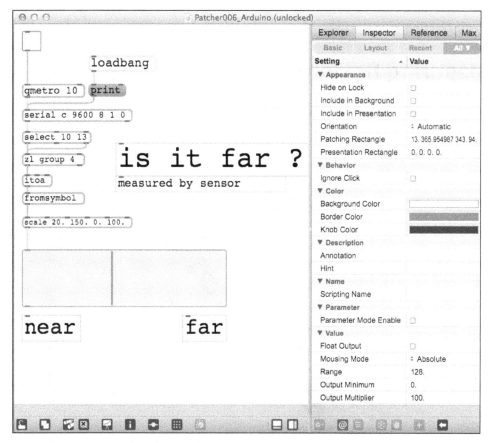

The inspector window showing the attributes and properties of the selected slider object

I won't describe all parameters, only the two at the bottom. In order to produce a relevant result, I had to scale the value coming from the `fromsymbol` object. I know the range of values transmitted by the Arduino (though this could require some personal verification), having calculated them from the Sharp datasheet. I considered this range as 20 to 150 cm. I mean a number between 20 and 150.

I took this range and compressed and translated it a bit, using the `scale` object, into a `0-to-100` range of float numbers. I chose the same range for my slider object. Doing that, the result displayed by the slider is coherent and represents the real value.

I didn't write any increment marks on the slider but only made two comments: `near` and `far`. It is a bit poetic in this world of numbers.

Let's check some other examples of sensors able to pop out continuous voltage variations.

Measuring flexion

Flexi sensors are also very much in use. Where the distance sensor is able to convert a measured distance into voltage, the flexi sensor measures flexion and provides a voltage.

Basically, the device flexion is related to a variable resistance able to make a voltage vary according to the amount of flexion.

A standard flexi sensor with two connectors only

A flexi sensor can be used for many purposes.

I like to use it to inform computer through Arduino about door position in digital installations I design. People wanted initially to know only about whether doors are open or closed, but I proposed to use a flexi and got very good information about the angle of openness.

The following figure illustrates how the sensor works:

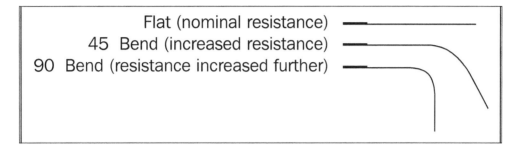

Now, I'm directly giving you the wiring schematic made again with Fritzing:

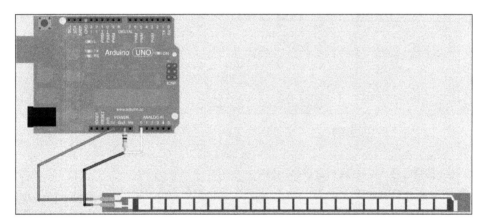

Flexi sensor connected to Arduino board with the pull-down resistor

I put a pull-down resistor. If you didn't read *Chapter 5, Sensing with Digital Inputs*, about pull-up and pull-down resistors, I suggest you to do that now.

Usually, I use resistors about 10K Ω and they work fine.

The circuit diagram is shown in the following figure:

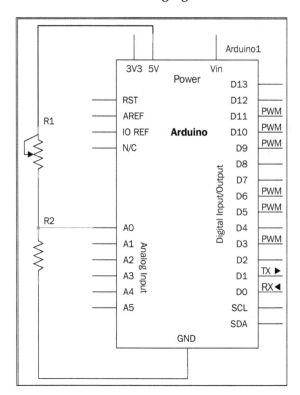

The flexi sensor and its pull-down resistor wired to the Arduino

Resistance calculations

For this project, I won't give you the code because it is very similar to the previous one, except for the calculation formulas. It is these resistance calculation formulas I'd like to discuss here.

What do we do if we don't have the graph the Sharp Co. was kind enough to include with their infrared sensor? We have to resort to some calculations.

Usually, the flexi sensor documentation provides resistance values for it when it is not bent and when it is bent at 90 degrees. Let's say some usual values of 10K Ω and 20K Ω, respectively.

What are the voltage values we can expect for these resistances values, including the pull-down one too?

Considering the electrical schematic, the voltage at the analog pin 0 is:

$$V_{A0} = \frac{R_{pulldown}}{R_{pulldown} + R_{flexi}} * 5V$$

If we choose the same resistance for the pull-down as the one for the flexi when it is not flexed, we can expect the voltage to behave according to this formula:

$$V_{A0} = \frac{10k}{10k + 20k} * 5V = 1.7V$$

Obviously, by using the same formula when it isn't bent, we can expect:

$$V_{A0} = \frac{10k}{10k + 10k} * 5V = 2.5V$$

This means we found our range of voltage values.

We can now convert that into digital 10-bit, encoded values, I mean the famous 0-to-1023 range of Arduino's ADC.

A small, easy calculation provides us with the values:

- `511` when the voltage is 2.5 (when the flexi isn't bent)
- `347` when the voltage is 1.7 (when the flexi is bent at around a 90-degree angle)

Because the voltage at Arduino's pin depends on the inverse of the resistance, we don't have a perfectly linear variation.

Experience tells me I can almost approximate this to a linear variation, and I used a scale function in Arduino firmware in order to map `[347,511]` to a simplerange of `[0,90]`. `map(value, fromLow, fromHigh, toLow, toHigh)` is the function to use here.

Do you remember the `scale` object in Max 6? `map()` works basically the same way, but for the Arduino. The statement here would be `map(347,511,90,0)`. This would give a fairly approximated value for the physical angle of bend.

The `map` function works in both directions and can map number segments going in the opposite direction. I guess you begin to see what steps to follow when you have to work with analog inputs on the Arduino.

Now, we are going to meet some other sensors.

Sensing almost everything

Whatever the physical parameter you want to measure, there's a sensor for it.

Here is a small list:

- Light color and light intensity
- Sound volume
- Radioactivity intensity
- Humidity
- Pressure
- Flexion
- Liquid level
- Compass and direction related to magnetic north
- Gas-specific detection
- Vibration intensity
- Acceleration on three axes (x, y, z)
- Temperature
- Distance
- Weight (different for a pure flexion sensor)

It isn't an exhaustive list, but it is quite complete.

Prices are really variable from a few dollars to $50 or $60. I found one of the cheaper Geiger counters for around $100. You can find a huge list of companies available on the Internet to buy sensors from in *Appendix G, List of Components' Distributors*.

Now, let's move further. How can we handle multiple analog sensors? The first answer is by wiring everything to many analog inputs of the Arduino. Let's check if we can be smarter than that.

Multiplexing with a CD4051 multiplexer/ demultiplexer

We are going to explore a technique called **multiplexing**. This is a major subchapter because we are going to learn how to make our real-life project more concrete, more real.

We often have many constraints in the real world. One can be the number of Arduinos available. This constraint can also come from having a computer that has only one USB port. Yes, this happens in real life, and I would be lying if I said you can have every connector you want, whenever you want, within the budget you want.

Imagine that you have to plug more than eight sensors to you Arduino's analog input. How would you do it?

We will learn to multiplex signals.

Multiplexing concepts

Multiplexing is quite common in the telecommunications world. Multiplexing defines techniques providing efficient ways to make multiple signals share a single medium.

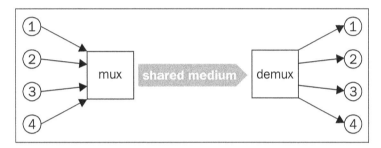

Basic multiplexing concept showing the shared medium

This technique provides a very helpful concept in which you only need one shared medium to bring many channels of information as we can see in the previous figure.

Of course, it involves multiplexing (named mux in the figure) and demultiplexing (demux) processes.

Let's dig into those processes a bit.

Multiple multiplexing/demultiplexing techniques

When we have to multiplex/demultiplex signals, we basically have to find a way to separate them using physical quantities that we can control.

I can list at least three types of multiplexing techniques:

- space-division multiplexing
- frequency-division multiplexing
- time-division multiplexing

Space-division multiplexing

This is the easiest to grasp.

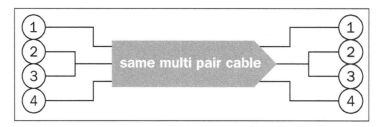

Space-division multiplexing physically agglomerates all wires into the same place

This concept is the basic phone network multiplexing in your flat, for instance.

Your phone wires go out, as those from your neighbors, and all those wires are joined into one shielded, big multipair cable containing, for instance, all phone wires for the whole building in which you live. This huge multipair cable goes into the street, and it is easier to catch it as a single global cable than if you had to catch each cable coming from your neighbors plus yours.

This concept is easily transposable to Wi-Fi communications. Indeed, some Wi-Fi routers today provide more than one Wi-Fi antenna. Each antenna would be able, for instance, to handle one Wi-Fi link. Every communication would be transmitted using the same medium: air transporting electromagnetic waves.

Frequency-division multiplexing

This type of multiplexing is very common in everything related to DSL and cable TV connections.

Service providers can (and do) provide more than one service on the same cable using this technique.

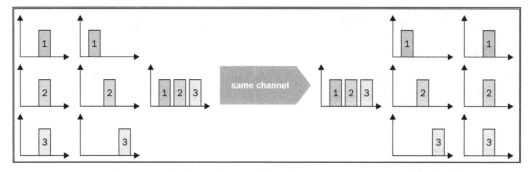

Frequency-division multiplexing plays with frequencies of transmission and bandwidths

Imagine the **1**, **2**, and **3** frequency bands on the figure would be three different services. 1 could be voice, 2 could be internet, and 3 TV. The reality isn't too far from this.

Of course, what we multiplex at one end, we have to demultiplex at the other in order to address our signals correctly. I wouldn't try to convert a TV modulated signal into voice, but I'm guessing it wouldn't be a very fruitful experience.

Time-division multiplexing

This is the case we are going to dig into the deepest because this is the one we are going to use with the Arduino to multiplex many signals.

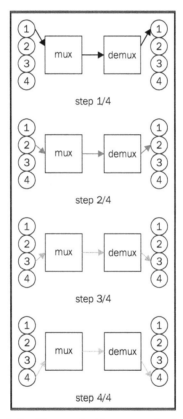

Time-division multiplexing illustrated with an example of one cycle of four steps

Sequentially, only one channel between the multiplexer and the demultiplexer is fully used for the first signal, then the second, and so on, until the last one.

This kind of system often involves a clock. This helps in setting the right cycle for each participant so they know at which step of communication we are. It is critical that we preserve the safety and integrity of communications.

Serial communications work like that, and for many reasons — even if you think you know them a lot after previous chapters — we'll dig a bit deeper into them in the next chapter.

Let's check how we can deal with eight sensors and only one analog input for our Arduino board.

The CD4051B analog multiplexer

The CD4051B analog multiplexer is a very cheap one and is very useful. It is basically an analog and digital multiplexer and demultiplexer. This doesn't mean you can use it as a multiplexer and a demultiplexer at the same time. You have to identify in what case you are and wire and design the code for this proper case. But it is always useful to have a couple of CD4051B devices.

Used as a multiplexer, you can connect, say eight potentiometers to the CD4051B and only one Arduino analog input, and you'll be able, by code, to read all 8 values.

Used as a demultiplexer, you could write to eight analog outputs by writing from only one Arduino pin. We'll talk about that a bit later in this book, when we approach the output pin and especially the **pulse-width modulation (PWM)** trick with LEDs.

What is an integrated circuit?

An **integrated circuit** (**IC**) is an electronic circuit miniaturized and all included in a small box of plastic. This is the simplest definition.

Basically, we cannot talk about integrated circuits without bringing to mind their small size. It is one of the more interesting features of IC.

The other one is what I am naming the **black box abstraction**. I also define it like the programming-like classes of the hardware world. Why? Because you don't have to know exactly how it works but only how you can use it. It means all the circuits inside don't really matter if the legs outside make sense for your own purpose.

Here are two among several type of IC packages:

- **Dual in-line package** (DIP, also named **DIL**)
- **Small outline** (SO)

You can find a useful guide at `http://how-to.wikia.com/wiki/Guide_to_IC_packages`.

The more commonly used of the two ICs are definitely DIPs. They are also called through-holes. We can easily manipulate and plug them into a breadboard or **printed circuit board (PCB)**.

SO requires more dexterity and finer tools.

Wiring the CD4051B IC?

The first question is about *what* it looks like? In this case, the answer is that it looks like a DIP package.

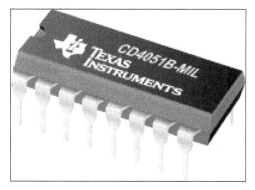

The CD4051B DIP case version

Here is the face of this nice little integrated circuit. The datasheet is easy to find on the Internet. Here is one by Texas Instruments:

`http://www.ti.com/lit/ds/symlink/cd4051b.pdf`

I redrew the global package in the next figure.

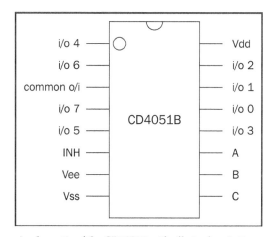

A schematic of the CD4051B with all pin descriptions

Identifying pin number 1

It is easy easy to find out which pin is pin number 1. As standard, there is a small circle engraved in front of one of the corner pins. This is the pin number 1.

There is also a small hole shaped as a half circle. When you place the IC with this half circle at the top (as shown on the previous figure), you know which pin number 1 is; the first pin next to pin number 1 is pin number 2, and so on, until the last pin of the left column which, in our case, is pin number 8. Then, continue with the pin opposite to the last one in the left column; this is pin number 9, and the next pin is pin number 10, and so on, until the top of the right column.

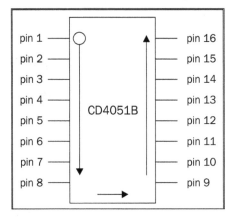

Numbering the pins of an IC

Of course, it would be much too simple if the first input was pin 1. The only real way you can know for sure is to check the specs.

Supplying the IC

The IC itself has to be supplied. This is to make it active but also, in some cases, to drive the current too.

- Vdd is the positive supply voltage pin. It has to be wired to the 5 V supply.
- Vee is the negative supply voltage pin. Here, we'll wire it to Ground.
- Vss is the ground pin, connected to Ground too.

Analog I/O series and the common O/I

Check the order of the I and the O in this title.

If you choose to use the CD4051B as a multiplexer, you'll have multiple analog inputs and one common output.

On the other hand, if you choose to use it as a demultiplexer, you'll have one common input and multiple analog outputs.

How does the selection/commutation work? Let's check the selector's digital pins, A, B, and C.

Selecting the digital pin

Now comes the most important part.

There are three pins, named A (pin 11), B (pin10), and C (pin 9), that have to be driven by digital pins of the Arduino. What? Aren't we in the analog inputs part? We totally are, but we'll introduce a new method of control using these three selected pins.

The multiplexing engine under the hood isn't that hard to understand.

Basically, we send some signal to make the CD4051B commute the inputs to the common output. If we wanted to use it as a demultiplexer, the three selected pins would have to be controlled exactly in the same way.

In the datasheet, I found a table of truth. What is that? It is just a table where we can check which A, B, and C combinations commute the inputs to the common output.

The following table describes the combination:

INPUT STATES				
INHIBIT	C	B	A	"ON CHANNEL(S)"
CD4051B				
0	0	0	0	0
0	0	0	1	1
0	0	1	0	2
0	0	1	1	3
0	1	0	0	4
0	1	0	0	5
0	1	1	0	6
0	1	1	1	7
1	X	X	X	None

The truth table for the CD4051B

In other words, it means that, if we write 1 to the digital output on Arduino corresponding to A, 1 to that corresponding to B and 0 to that corresponding to C, the commuted input would be the third channel.

Of course, there is something good in this. If you *read* the binary number corresponding to the inputs on C, B, and A (in that order), you'll have a nice surprise; it will be equivalent to the decimal number of the input pin commuted by the common output.

Indeed, 0 0 0 in binary equals 0 in decimal. Refer the table for the binary values of decimal numbers:

0 0 0	0
0 0 1	1
0 1 0	2
0 1 1	3
1 0 0	4
1 0 1	5
1 1 0	6
1 1 1	7

Here is how we could wire things:

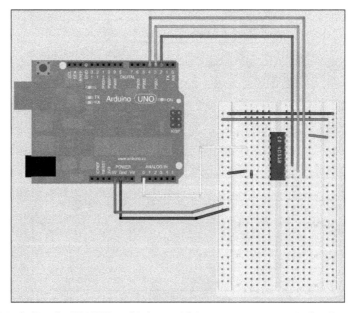

The circuit including the CD4051B multiplexer with its common output wired to the analog pin 0

And the following figure is the electrical diagram:

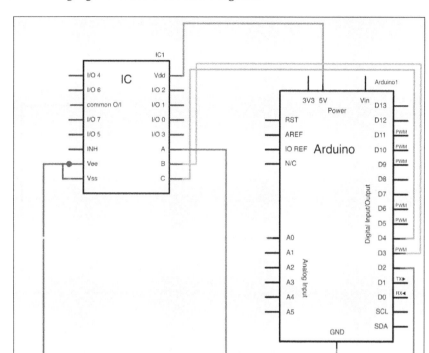

The electrical diagram

All devices we'd like to read with this system should be wired to I/O 0, 1, 2, and so on, on the CD4051B.

Considering what we know about table of truth and how the device works, if we want to read sequentially all pins from 0 to 7, we will have to make a loop containing both types of statements:

- One for commuting the multiplexer
- One for reading the Arduino analog input 0

The source code would look like this (you can find it in the `Chapter6/analogMuxReader` folder):

```
int muxOutputPin = 0 ;  // pin connected to the common output of the
CD4051B
int devicesNumber = 8 ; // number of device // BE CAREFUL, plug them
from 0
```

```
int controlPinA = 2 ;    // pin connected to the select pin A of the
CD4051B
int controlPinB = 3 ;    // pin connected to the select pin B of the
CD4051B
int controlPinC = 4 ;    // pin connected to the select pin C of the
CD4051B

int currentInput = 0 ;   // hold the current analog input commuted o
the common output of the CD4051B

void setup() {
  Serial.begin(9600);

  // setting up all 3 digital pins related to selectors A, B and C as
outputs
  pinMode(controlPinA, OUTPUT);
  pinMode(controlPinB, OUTPUT);
  pinMode(controlPinC, OUTPUT);
}

void loop(){
  for (currentInput = 0 ; currentInput < devicesNumber - 1 ;
currentInput++)
  {
    // selecting the inputs that is commuted to the common output of
the CD4051B
    digitalWrite(controlPinA, bitRead(currentInput,0));
    digitalWrite(controlPinB, bitRead(currentInput,1));
    digitalWrite(controlPinC, bitRead(currentInput,2));

    // reading and storing the value of the currentInput
    Serial.println(analogRead(muxOutputPin)) ;
  }
}
```

After you've defined all the variables, we set up the serial port in `setup()` and also the three pins related to the selector pin of the CD4051B as outputs. Then, in each cycle, I first select the commuted input by either driving the current or not to pins A, B, and C of the CD4051B. I'm using a nested function in my statement in order to save some rows.

`bitRead(number,n)` is a new function able to return the *nth* bit of a number. It is the perfect function for us in our case.

We make a loop over the input commuted from 0 to 7, more precisely to `devicesNumber - 1`.

By writing those bits to pins A, B, and C of the CD4051B device, it selects the analog input at each turn and pops the value read at the serial port for further processing in Processing or Max 6 or whatever software you want to use.

Summary

In this chapter, we learnt at least how to approach a very powerful graphical framework environment named Max 6. We'll use it in several further examples in this book as we continue to use Processing too.

We learnt some reflexes for when we want to handle sensors providing continuous voltage variations to our Arduino analog inputs.

Then, we also discovered a very important technique, the multiplexing/ demultiplexing.

We are going to talk about it in the next chapter about serial communication. We'll dig deeper into this type of communication now that we have used a lot of time already.

7
Talking over Serial

We already saw that using Arduino is all about talking and sharing signals. Indeed, from the most basic component in Arduino, reacting to some physical world values by changing its environment and propagating the change as a basic message to its neighbors, to the now classic serial communication, electronic entities are talking among themselves and to us.

As with the many concepts in this book, we have already used serial communication and the underlying Serial protocol a couple of times as a black-boxed tool, that is, a tool I have introduced but not explained.

We are going to dive into it in this small chapter. We will discover that serial communication is used not only for machine-to-human communication but also for "component-to-component" discussions inside machines. By components, I mean small systems, and I could use the term peripheral to describe them.

Serial communication

Typically, serial communication in computer science and telecommunications is a type of communication where data is sent one bit at a time over a communication bus.

Nowadays, we can see serial communication all around us, and often we don't even realize this. J The "S" in the **USB** acronym means Serial (USB is **Universal Serial Bus**), and represents the underlying serial communication bus used by every higher protocol.

Let's dig into that right now.

Serial and parallel communication

Serial communication is often defined by its opposite form of communication, **parallel communication**, where several bits of data are sent out over a link made by several parallel channels at the same time. Look at the following figure:

Basic, unidirectional serial communication between a speaker and a listener

Now let's compare this to a parallel case:

Basic, unidirectional parallel communication between a speaker and a listener

In these two figures, a speaker is sending the following data byte: 0 1 1 0 0 0 1 1. These eight bits of data are sent sequentially over one channel in the case where serial communication has been used, and simultaneously over eight different channels in the case where parallel communication has been used.

Right from small-distance to long-distance communications, even if the parallel approach seems faster at first glance because more than one bit of data is sent at the same time during a clock cycle, serial communication has progressively outperformed other forms of communication.

The first reason for this is the number of wires involved. For example, the parallel approach used in our small example requires eight channels to drive our eight bits of data at the same time, while the serial requires only one. We'll discuss what a channel is very soon, but with one wire, the ratio 1:8 would save us money if we were to use serial communication.

The second major reason is the fact that we finally achieved to make serial communication very fast. This has been achieved due to the following:

- Firstly, **propagation time** is easier to handle with a smaller number of wires
- Secondly, **crosstalk** is less with fewer channels than with a higher density of channels such as those found in parallel links
- Thirdly, because there are fewer wires involved, we can save space (and money) and often use this saved space to shield our wires better

Nowadays, serial communication bandwidths range from several megabits per second to more than 1 terabit per second (which means 1,000 gigabits per second), and a lot of media can be used from wire-driven fibers to wireless, and from copper cables to optical fibers. As you might suspect, there are many serial protocols that are used.

Types and characteristics of serial communications

Whether it be synchronism or duplex mode or bus or peering, serial communication can be defined differently, and we have to dig into that point here.

Synchronous or asynchronous

Serial communication can either be synchronous or not.

Synchronous communication involves a clock, which we can call a master clock, that keeps a reference time for all the participants of the communication. The first example that comes to mind is phone communication.

Asynchronous communication doesn't require that the clock's data be sent over the serial channel(s); this makes it easier to communicate but it can lead to some issues with understandability at times. Mailing and texting are asynchronous types of communication.

Duplex mode

The duplex mode is a particular characteristic of a communication channel. It can be:

- **Simplex**: Unidirectional only (data is passed only in one direction, between two points)
- **Half-duplex** : Bidirectional, but only in one direction at the same time
- **Full-duplex** : Bidirectional simultaneously

Half-duplex is obviously more useful than simplex, but it has to run a collision detection and retransmission process. Indeed, when you are talking to your friend, you are also sharing the same media (the room and air inside the room that carries vibrations from your mouth to his ears), and if you are talking at the same time, usually one checks that and stops and tells the other to repeat.

Full-duplex requires more channels. That way no collisions occur and all the collision detection and retransmission processes can be dropped. The detection of other errors and fixing is still involved, but usually it is much easier.

Peering and bus

In a **peering** system, the speakers are linked to listeners either physically or logically. There is no master, and these kinds of interfaces are most often asynchronous.

In a **bus**, they will all get connected physically at some point and some logical commutations will occur.

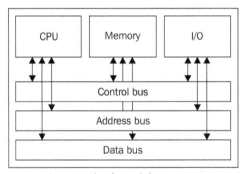

An example of a multibus system

Master and slave buses

In master/slave buses, one device is the master and the others are the slaves, and this usually involves synchronism where the master participant generates the timing clock.

The main difficulty with serial communication is to avoid collisions and misunderstandings.

There are a lot of solutions that can be implemented to solve these problems, such as using multiple physical link types and specific preexisting communication protocols. Let's check some of these, and especially those we can use with Arduino of course.

Data encoding

The most important things to define when we use serial protocols for our communication are as follows:

- The word length in bits

- Whether a stop bit is present or not (defines a blank moment in time)

- Whether a parity bit is present or not (defines the simplest, error-detecting, code-based solution)

Indeed, especially in asynchronous communication, how could a listener know where a word begins or ends without these properties? Usually, we hardcode this behavior in both the participants' brains in order to be sure we have a valid communication protocol.

In the first figure of this chapter, I sent eight bits of data over the channel. This equals 1 byte.

We often write the types of serial communication as `<word length><parity><stop>`. For instance, 8 bit without parity but one stop is written as `8N1`.

I won't describe the parity bit completely, but you should know that it is basically a checksum. Using this concept, we transmit a word and checksum, after which we verify the binary sum of all the bits in my received word. In this way, the listener can check the integrity of the words that were received quite easily, but in a very primitive way. An error can occur, but this is the cheapest way; it can avoid a lot of errors and is statistically right.

A global frame of data with the `8N1` type serial communication contains 10 bits:

- One start bit

- Eight bits for each characters

- One stop bit

Indeed, only 80 percent of the data sent is the real payload. We are always trying to reduce the amount of flow control data that is sent because it can save bandwidth and ultimately time.

Multiple serial interfaces

I won't describe all the serial protocols, but I'd like to talk about some important ones, and sort them into families.

The powerful Morse code telegraphy ancestor

I give you one of the oldest Serial protocols: the Morse code telegraphy protocol. Telecommunications operators have been using this one since the second half of the 19th century.

I have to say that Samuel F. B. Morse was not only an inventor but also an accomplished artist and painter. It is important to mention this here because I'm really convinced that art and technology are finally one and the same thing that we used to see with two different points of view. I could quote more artist/inventor persons but I guess it would be a bit off topic.

By sending long and short pulses separated by blanks, Morse's operators can send words, sentences, and information. This can happen over multiple types of media, such as:

- Wires (electrical pulses)
- Air (electromagnetic wave carriers, light, sounds)

It can be sorted into a peered, half-duplex, and asynchronous communication system.

There are some rules about the duration of pulses ranging from long to short to blank, but this remains asynchronous because there isn't really a clock shared between both participants.

The famous RS-232

RS-232 is a common interface that you will find on all personal computers. It defines a complete standard for electrical to physical (and electrical to mechanical) characteristics, such as connection hardware, pins, and signal names. RS-232 was introduced in 1962 and is still widely used. This point-to-point interface can drive data up to 20 Kbps (kilobit per second = 20,000 bits per second) for moderate distances. Even though it isn't specified in the standard, we will usually find instances where the speed is greater than 115.2 Kbps on short and shielded wires.

I myself use cables that are 20 meters long with sensors that transmit their data over serial to Arduino for different installations. Some friends use cables that are 50 meters long, but I don't do that and prefer other solutions such as Ethernet.

From 25 wires to 3

If the standard defines a 25-pin connector and link, we can reduce this huge number required for multiple hardware flow control, error detections, and more to only three wires:

- Transmit data (usually written as TX)
- Receive data (usually written as RX)
- Ground

The connector with 25 pins/wires is named DB25 and has been used a lot, for peripherals such as printers. There is another type of connector named DB9 with 9 pins/wires only. This is a variant that omits more wires than DB25. This DB9 has been used a lot for connecting mouse devices.

But how can we omit a large number of wires/signals and keep the serial communication working well? Basically, as with many standards, it has been designed to fit a lot of use cases. For instance, in the full version of DB25, there are pins 8 and 22 that are dedicated to phone lines: the first one is the **Data Carrier Detect** and the second one is the **Ring Indicator**. The signal sent over pins 4 and 5 is used for a handshake between the participants.

In this standard, pin 7 is the common ground and 2 and 3 are respectively TX and RX. With only these three, we can make our serial asynchronous communication correctly.

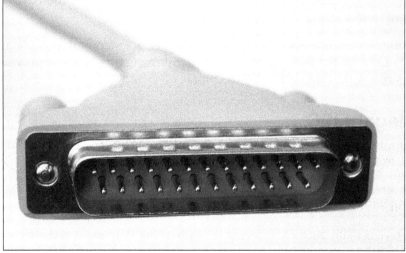

The DB25 connector

The DB9 connector

Our precious Arduino provides this three-wire serial alternative. Of course, each type of board doesn't provide the same number of serial interfaces, but the principle remains the same: a serial interface based on three-wire is available.

Arduino Uno and Leonardo provide the three wires TX, RX, and ground, while freshly released Arduino Mega 2560 and Arduino Due (`http://arduino.cc/en/Main/ArduinoBoardDue`) provide four different serial communication interface names right from RX0 and TX0 to RX3 and TX3.

We are going to describe another type of serial interface standard, and we'll come back to RS-232 with the famous integrated circuit made by FTDI that provides a very efficient way to convert RS-232 to USB.

The elegant I2C

The I2C multimaster serial single-ended computer bus has been designed by Philips and requires a license for any hardware implementations.

One of its advantages is the fact that it uses only two wires: **SDA (Serial Data Line)** with a 7-bit addressing system and **SCL (Serial Clock Line)**.

This interface is really nice considering its addressing system. In order to use it, we have to build the two wire-based bus from Arduino, which is the master here.

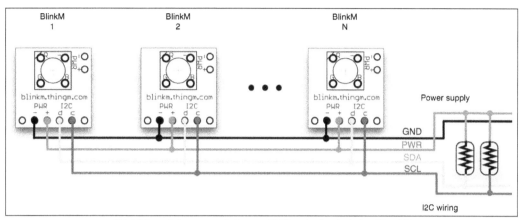

BlinkM modules connected as an I2C bus to the Arduino Uno R3

In order to know which pins have to be used for each Arduino board, you can directly check the information at `http://www.arduino.cc/en/Reference/Wire`.

BlinkM modules (`http://thingm.com/products/blinkm`) are RGB LED modules with a small form factor that are quite easy to manipulate on I2C buses. I also used it a lot to more or less control big LCDs with Arduino.

This is also the page of the `Wire` library for Arduino. Nowadays, this library is included with the Arduino core. Considering the complexity of the standard, the cost increases when you have a lot of elements on the buses. Because of its two wires and the precision of data integrity, this is still an elegant solution for short-distance and intermittent communication inside the same box. The **Two Wire Interface** (TWI) is principally the same standard as I2C. It was known by another name when the patents on I2C were still running.

I2C has been the base for many other interface protocols, such as VESA DDC (a digital link between screens and graphical card), SMBus by Intel, and some others.

The synchronous SPI

SPI stands for **Serial Peripheral Interface**, which has been developed by Motorola. It uses the following four wires:

- **SCLK**: This is the serial clock driven by the master
- **MOSI**: This is the master output / slave input driven by the master
- **MISO**: This is the master input / slave output driven by the master
- **SS**: This is the slave-selection wire

It is very useful in point-to-point communication where there is only one master and one slave, even if we find many applications with more than one slave on the SPI bus.

Since SPI is a full-duplex, mode-based interface, we can achieve higher data rates than we can with I2C. It is often used for communication between a coder/decoder and a digital signal processor; this communication consists of sending samples in and out at the same time. SPI lacking device addressing is also a huge advantage as it makes it much lighter and thus faster in case you don't need this feature. Indeed, I2C and SPI are really complementary to each other depending on what you want to achieve.

There is information available online regarding SPI in the Arduino boards (`http://arduino.cc/en/Reference/SPI`), but you have to know that we can easily use any digital pins as one of the four wires included in SPI.

I personally have often used it in projects involving a lot of shift registers that are all daisy-chained to have a lot of inputs and/or outputs with Arduino Uno and even Arduino Mega, this latter offering more outputs and inputs natively.

We'll describe the use of shift registers in the next chapter when I show you how to multiplex outputs quite easily with some smart and, ultimately, very simple integrated circuits linked to Arduino through SPI.

The omnipresent USB

USB is the Universal Serial Bus standard. This is probably the one you use the most.

The main advantage of this standard is the Plug and Play feature of USB devices. You can plug and unplug devices without restarting your computer.

USB has been designed to standardize the connection of a wide variety of computer peripherals, including the following:

- Audio (speaker, microphone, sound card, MIDI)
- Communications (modem, Wi-Fi, and Ethernet)
- Human interface device (HID, keyboard, mouse, joystick)
- Image and video (webcam, scanner)
- Printer
- Mass storage (flash drive, memory card, drive)
- Wireless (infrared)

And there are many more types too. The standard is actually Version 3.0. A USB bus can contain up to 127 peripherals and can supply a maximum of 500 to 900 mA for general devices.

USB system design

The architecture of USB is an asymmetrical topology consisting of one host and a multitude of downstream USB ports and multiple peripheral devices connected in a tiered-star topology.

USB hubs can be included in the tiers allowing branching up to five tier levels. This results in a tree topology. This is why you can stack hubs on hubs.

Device classes provide a way of having an adaptable and device-independent host to support new devices. An ID that the host can recognize defines each class. You can find all the approved classes at `http://www.usb.org/developers/devclass` on the official USB standard website.

USB connectors and cables

A USB standard plug contains four wires (`http://en.wikipedia.org/wiki/Universal_Serial_Bus`):

- Vcc (+5 V)
- Data-
- Data+
- Ground

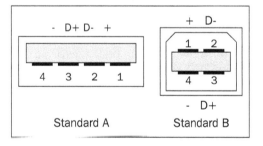

The USB standard A and B plugs

Cables are shielded; their usual maximal lengths are around two to five meters. I already used a 12-meter cable for the USB port. It worked totally fine with a cable that I myself soldered in an electromagnetic-safe environment, I mean, in a place where my cable was alone behind a wall and not mixed with a lot of other cables, especially the ones supplying power.

There are some other types of plug that are somewhat bigger, but the requirement of having at least four wires remains the same.

FTDI IC converting RS-232 to USB

Except for some versions, such as the Arduino Pro Mini, Arduino boards provide a USB connector, as you already know and have used.

This provides the basic power supply feature for a computer or the hubs connected to a computer, and it is used for communication too.

The FTDI integrated circuit EEPROM named FT232 provides a way of converting USB into an RS-232 interface. This is why we can use the serial communication features of the Arduino boards over USB without the need for an external serial port interface from the Arduino pins related to serial communication, which are TX and RX. New boards include an Atmega16U2 that provides serial communication features.

Indeed, as soon as you connect your Arduino board to a computer, you will have a serial communication feature available. We already used it with:

- Arduino IDE (Serial Monitor)
- Processing (with the serial library)
- Max 6 (with the serial object)

I guess you also recall that we weren't able to use the Serial Monitor while using Max 6's serial object polling feature.

Do you understand why now? Only one point-to-point link can be active at the same time on the wires and in the virtual world of computers. It's the same for physical links, too. I warned you not to use the digital pins 0 and 1 as soon as you needed to use serial communication with the Arduino board, especially the Diecimilla version. These pins are directly connected to the corresponding RX and TX pins of the FTDI USB-to-TTL serial chip.

 If you use serial communication over the USB feature, you have to avoid using the digital pins 0 and 1.

Summary

In this chapter, we talked about serial communication. This is a very common mode of communication both inside and between electronic devices. This chapter is also a nice introduction to other communication protocol in general, and I'm sure that you are now ready to understand more advanced features.

In the next chapter, we'll use some of the different types of serial protocol that were introduced here. In particular, we are going to talk about Arduino outputs; this means that not only will we be able to add feedback and reactions to our Arduino boards, considering behavior pattern designs such as stimulus and response for deterministic ways, but we will also see more chaotic behaviors such as those including constrained chance, for instance.

8
Designing Visual Output Feedback

Interaction is everything about control and feedback. You control a system by performing actions upon it. You can even modify it. The system gives you feedback by providing useful information about what it does when you modify it.

In the previous chapter, we learned more about us controlling Arduino than Arduino giving us feedback. For instance, we used buttons and knobs to send data to Arduino, making it working for us. Of course, there are a lot of point of view, and we can easily consider controlling an LED and giving feedback to Arduino. But usually, we talk about feedback when we want to qualify a return of information from the system to us.

Arkalgud Ramaprasad, Professor at the Department of Information and Decision Sciences at the College of Business Administration, University of Illinois, Chicago, defines feedback as follows:

> *"Information about the gap between the actual level and the reference level of a system parameter which is used to alter the gap in some way."*

We already talked about some visual output in *Chapter 5, Sensing Digital Inputs*, when we tried to visualize the result of our button push events. This visual rendering resulting from our push events was feedback.

We are now going to talk about the design of visual feedback systems based especially on LEDs driven by the Arduino board. LEDs are the easiest systems with which to provide visual feedback from Arduino.

We are going to learn about the following:

- How to use basic monochromatic LEDs
- How to make LED matrices and how to multiplex LEDs
- How to use RGB LEDs

We will finish the chapter by introducing the LCD display device.

Using LEDs

LEDs can be monochromatic or polychromatic. Indeed, there are many types of LEDs. Before going though some examples, let's discover some of these LED types.

Different types of LEDs

Usually, LEDs are used both to block the current coming from a line to its cathode leg and to give light feedback when the current goes into its anode:

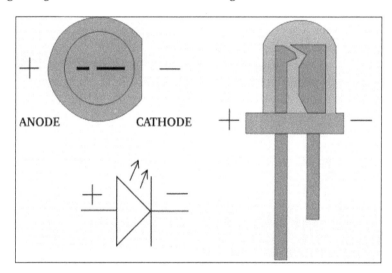

The different models that we can find are as follows:

- Basic LEDs
- **OLED (Organic LED** made by layering the organic semi-conductor part)
- **AMOLED (Active Matrix OLED** provides a high density of pixels for big size screens)
- **FOLED (Flexible OLED)**

We will only talk about basic LEDs here. By the term "basic", I mean an LED with discrete components like the one in the preceding image.

The package can vary from two-legged components with a molded epoxy-like lens at the top, to surface components that provide many connectors, as shown in the following screenshot:

We can also sort them, using their light's color characteristics, into:

- Monochromatic LEDs
- Polychromatic LEDs

In each case, the visible color of an LED is given by the color of the molded epoxy cap; the LED itself emits the same wavelength.

Monochromatic LEDS

Monochromatic LEDs emit one color only.

The most usual monochromatic LEDs emit constant colors at each voltage need.

Polychromatic LEDs

Polychromatic LEDs can emit more than one color, depending on several parameters such as voltage but also depending on the leg fed with current in case of an LED with more than one leg.

The most important characteristic here is controllability. Polychromatic LEDs have to be easily controllable. This means that we should be able to control each color by switching it on or off.

Here is a classic RGB LED with common cathode and three different anodes:

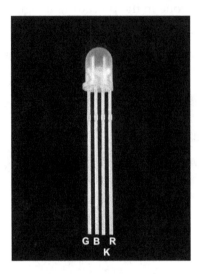

This type of LED is the way to go with our Arduino stuff. They aren't expensive (around 1.2 Euros per 100 LEDs), considering the fact that we can control them easily and produce a very huge range of colors with them.

We are going to understand how we can deal with multiple LEDs and also polychromatic RGB LEDs in the following pages.

Remembering the Hello LED example

In Hello LED, we made an LED blink for 250 ms of every 1000 ms that pass. Let's see its schematic view once again to maintain the flow of your reading:

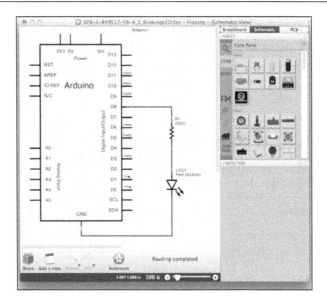

The code for Hello LED is as follows:

```
// Pin 8 is the one connected to our pretty LED
int ledPin = 8;                  // ledPin is an integer variable
initialized at 8

void setup() {
  pinMode(ledPin, OUTPUT);      // initialize the digital pin as an
output
}

// --------- the loop routine runs forever
void loop() {
  digitalWrite(ledPin, HIGH);   // turn the LED on
  delay(250);                   // wait for 250ms in the current state
  digitalWrite(ledPin, LOW);    // turn the LED off
  delay(1000);                  // wait for 1s in the current state
}
```

Intuitively, in the next examples, we are going to try using more than one LED, playing with both monochromatic and polychromatic LEDs.

Multiple monochromatic LEDs

Since we are talking about feedback here, and not just pure output, we will build a small example showing you how to deal with multiple buttons and multiple LEDs. Don't worry if you are totally unable to understand this right now; just continue reading.

Two buttons and two LEDs

We already spoke about playing with multiple buttons in *Chapter 5, Sensing Digital Inputs*. Let's build a new circuit now.

Here are the schematics:

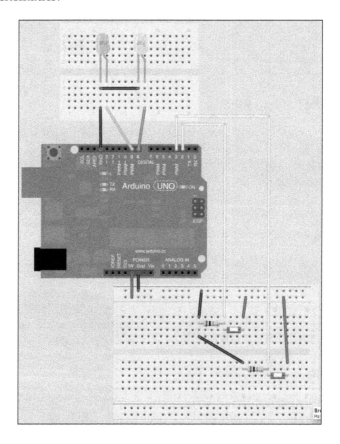

It's preferable to continue drawing the electric diagram related for each schematic.

Basically, the multiple buttons example from *Chapter 5, Sensing Digital Inputs*; however, we have removed one button and added two LEDs instead.

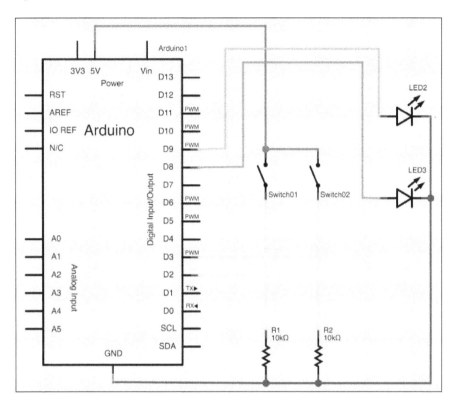

As you know, the digital pins of Arduino can be used as inputs or outputs. We can see that two switches are connected on one side to a 5 V Arduino pin and on the other side to the digital pins 2 and 3, with one pull-down resistor related to each of those latter pins, sinking the current to Arduino's ground pin.

We can also see that an LED is connected each to digital pin 8 and 9 on one side; both are connected to Arduino's ground pin.

Nothing's really incredible about that.

Before you design a dedicated firmware, you need to briefly cover something very important: coupling. It is a must to know for any interface design; more widely for interaction design.

Control and feedback coupling in interaction design

This section is considered a subchapter for two main reasons:

- Firstly, it sounds great and is key to keeping the motivation groove on
- Secondly, this part is the key for all your future human-machine interface design

As you already know, Arduino (thanks to its firmware) links the control and feedback sides. It is really important to keep this in mind.

Whatever the type of the external system may be, it is often considered as human from the Arduino point of view. As soon as you want to design an interaction system, you will have to deal with that.

We can summarize this concept with a very simple schematic in order to fix things in the mind.

Indeed, you have to understand that the firmware we are about to design will create a control-feedback coupling.

A **control/feedback coupling** is a set of rules that define how a system behaves when it receives orders from us and how it reacts by giving us (or not) feedback.

This hard-coded set of rules is very important to understand.

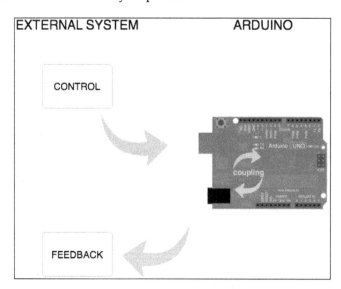

But, imagine that you want to control another system with Arduino. In that case, you may want to make the coupling outside Arduino itself.

See the second figure **EXTERNAL SYSTEM 2**, where I put the coupling outside Arduino. Usually, **EXTERNAL SYSTEM 1** is us and **EXTERNAL SYSTEM 2** is a computer:

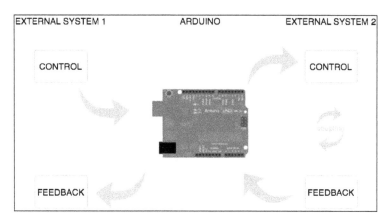

We can now quote a real-life example. As with many users of interfaces and remote controllers, I like and I need to control complex software on my computer with minimalistic hardware gears.

I like the minimalistic and open source **Monome interface** (http://monome.org) designed by Brian Crabtree. I used it a lot, and still use it sometimes. It is basically a matrix of LEDs and buttons. The amazing trick under the hood is that there is NO coupling inside the gear.

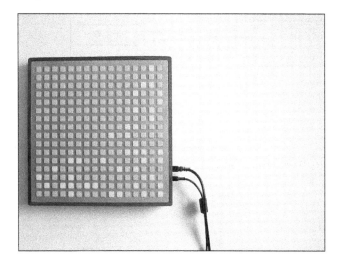

The preceding image is of Monome 256 by Brian Crabtree and its very well-made wooden case.

If it isn't directly written like that in all the docs, I would like to define it to my friends and students like this: "The Monome concept is the most minimalistic interface you'll ever need because it only provides a way of controlling LEDs; beside of that, you have many buttons, but there are no logical or physical links between buttons and LEDs."

If Monome doesn't provide a real, already made coupling between buttons and LEDs, it's because it would be very restrictive and would even remove all the creativity!

Since there is a very raw and efficient protocol designed (`http://monome.org/data/monome256_protocol.txt`) to especially control LEDs and read buttons pushes, we are ourselves able to create and design our own coupling. Monome is also provided with the **Monome Serial Router**, which is a very small application that basically translates the raw protocol into **OSC** (`http://archive.cnmat.berkeley.edu/OpenSoundControl/`) or **MIDI** (`http://www.midi.org/`). We will discuss them in later sections of this chapter. These are very common in multimedia interaction design; OSC can be transported over networks, while MIDI is very suited for links between music-related equipment such as sequencers and synthesizers.

This short digression wouldn't be complete without another schematic about the Monome.

Check it and let's learn more about it after that:

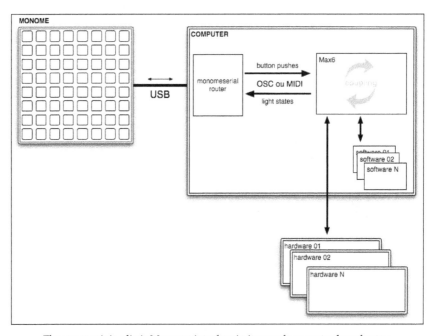

The smart minimalistic Monome interface in its usual computer-based setup

Here is a schematic of the Monome 64 interface, in that usual computer-based setup inside of which the coupling occurs. This is the real setup that I used on stage for a music performance many times (https://vimeo.com/20110773).

I designed a specific coupling inside Max 6, translating specific messages from/to the Monome itself, but from/to the software too, especially Ableton Live (https://www.ableton.com).

This is a very powerful system that controls things and provides feedback with which you can basically build your coupling from the ground up and transform your raw and minimalistic interface into whatever you need.

This was a small part of a more global monologue about interaction design.

Let's build this coupling firmware right now, and see how we can couple controls and feedback into a basic sample code.

The coupling firmware

Here, we only use the Arduino switches and LEDs and no computer actually.

Let's design a basic firmware, including coupling, based on this pseudocode:

- If I push switch 1, LED 1 is switched on, and if I release it, LED 1 is switched off
- If I push switch 2, LED 2 is switched on, and if I release it, LED 2 is switched off

In order to manipulate new elements and ideas, we are going to use a library named Bounce. It provides an easy way to debounce digital pin inputs. We already spoke about debouncing in the *Understanding the debounce concept* section of *Chapter 5, Sensing Digital Inputs*. Reminding you of that a bit: if no button absorbs the bounce totally when you push it, we can smoothen things and filter the non-desired harsh value jumps by using software.

You can find instructions about the Bounce library at http://arduino.cc/playground/Code/Bounce.

Let's check that piece of code:

```
#include <Bounce.h>   // include the (magic) Bounce library

#define BUTTON01 2    // pin of the button #1
#define BUTTON02 3    // pin of the button #2

#define LED01 8       // pin of the button #1
```

```
#define LED02 9          // pin of the button #2

// let's instantiate the 2 debouncers with a debounce time of 7 ms
Bounce bouncer_button01 = Bounce (BUTTON01, 7);
Bounce bouncer_button02 = Bounce (BUTTON02, 7);


void setup() {

  pinMode(BUTTON01, INPUT); // the switch pin 2 is setup as an input
  pinMode(BUTTON02, INPUT); // the switch pin 3 is setup as an input

  pinMode(LED01, OUTPUT);   // the switch pin 8 is setup as an output
  pinMode(LED02, OUTPUT);   // the switch pin 9 is setup as an output
}

void loop(){

  // let's update the two debouncers
  bouncer_button01.update();
  bouncer_button02.update();

  // let's read each button state, debounced!
  int button01_state = bouncer_button01.read();
  int button02_state = bouncer_button02.read();

  // let's test each button state and switch leds on or off
  if ( button01_state == HIGH ) digitalWrite(LED01, HIGH);
  else digitalWrite(LED01, LOW);

  if ( button02_state == HIGH ) digitalWrite(LED02, HIGH);
  else digitalWrite(LED02, LOW);
}
```

You can find it in the Chapter08/feedbacks_2x2/ folder.

This code includes the Bounce header file, that is, the Bounce library, at the beginning.

Then I defined four constants according to the digital input and output pins, where we put switches and LEDs in the circuit.

The Bounce library requires to instantiate each debouncer, as follows:

```
Bounce bouncer_button01 = Bounce (BUTTON01, 7);
Bounce bouncer_button02 = Bounce (BUTTON02, 7);
```

I chose a debounce time of 7 ms. This means, if you remember correctly, that two value changes occurring (voluntarily or non-voluntarily) very fast in a time interval of less than 7ms wouldn't be considered by the system, avoiding strange and uncanny bouncing results.

The `setup()` block isn't really difficult, it only defines digital pins as inputs for buttons and outputs for LEDs (please remember that digital pins can be both and that you have to choose at some point).

`loop()` begins by the update of both debouncers, after which we read each debounced button state value.

At last, we handle the LED controls, depending on the button states. Where does the coupling occur? Of course, at this very last step. We couple our control (buttons pushed) to our feedback (LED lights) in that firmware. Let's upload and test it.

More LEDs?

We basically just saw how to attach more than one LED to our Arduino. Of course, we could do the very same way with more than two LEDs. You can find code handling six LEDs and six switches in the `Chapter05/feedbacks_6x6/` folder.

But hey, I have a question for you: how would you handle more LEDs with an Arduino Uno? Please don't answer that by saying "I'll buy an Arduino MEGA" because then I would ask you how you would handle more than 50 LEDs.

The right answer is **multiplexing**. Let's check how we can handle a lot of LEDs.

Multiplexing LEDs

The concept of multiplexing is an interesting and efficient one. It is the key to having a bunch of peripherals connected to our Arduino boards.

Multiplexing provides a way to use few I/O pins on the board while using a lot of external components. The link between Arduino and these external components is made by using a multiplexer/demultiplexer (also shortened to mux/demux).

We spoke about input multiplexing in *Chapter 6, Playing with Analog Inputs*.

We are going to use the 74HC595 component here. Its datasheet can be found at `http://www.nxp.com/documents/data_sheet/74HC_HCT595.pdf`.

This component is an 8-bit serial-in / serial-or-parallel-out. This means it is controlled through a serial interface, basically using three pins with Arduino and can drive with eight of its pins.

I'm going to show you how you can control eight LEDs with only three pins of your Arduino. Since Arduino Uno contains 12 digital usable pins (I'm not taking 0 and 1, as usual), we can easily imagine using 4 x 75HC595 to control 4 x 8 = 32 monochromatic LEDs with this system. I'll provide the code to do that as well.

Connecting 75HC595 to Arduino and LEDs

As we learnt together with the CD4051 and the multiplexing of analog inputs, we are going to wire the chip to a 75HC595 shift register in order to mux/demux eight digital output pins. Let's check the wiring:

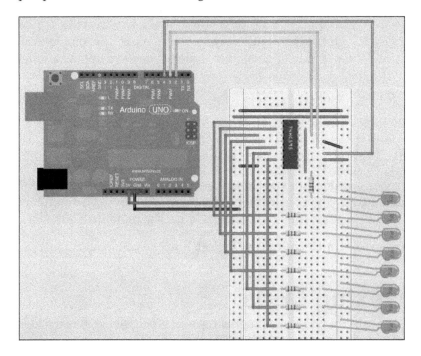

We have the Arduino supplying power to the breadboard. Each resistor provides 220 ohms resistance.

The 75HC595 grabs the GND and 5 V potential for its own supply and configuration.

Basically, 74HC595 needs to be connected through pins 11, 12, and 14 in order to be controlled by a serial protocol handled here by Arduino.

Let's check 74HC595 itself:

- Pins 8 and 16 are used for internal power supply.

- Pin 10 is named **Master Reset**, and in order to activate it, you have to connect this pin to the ground. That is the reason why, in normal operational states of work, we drive it to 5 V.

- Pin 13 is the output enable input pin and has to be kept active in order to make the whole device output currents. Connecting it to the ground does this.

- Pin 11 is the shift register clock input.

- Pin 12 is the storage register clock input, also named **Latch**.

- Pin 14 is the serial data input.

- Pin 15 and pins 1 to 7 are the output pins.

Our small and inexpensive serial link to the Arduino, handled by pins 11, 12 and 14, provides an easy way to control and basically load eight bits into the device. We can cycle over the eight bits and send them serially to the device that stores them in its registers.

These types of devices are usually referred to as **Shift Registers** we shift bits from 0 to 7 while loading them.

Then, each state is outputted to the correct output from Q0 to Q7, transposing the previously transmitted states over serial.

This is a direct illustration of the serial-to-parallel conversion that we talked about in the previous chapter. We had a data flow coming sequentially, retained until the register is globally loaded, then pushing this to many output pins.

Now, let's visualize the wiring diagram:

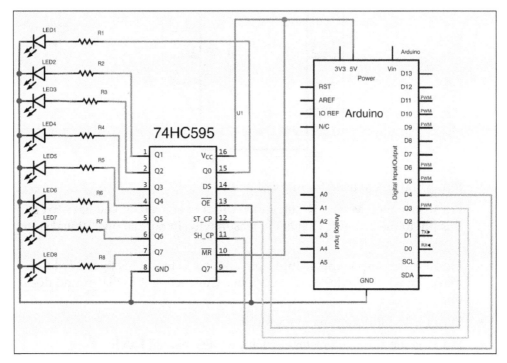

An eight-LED array with resistors wired to the 74HC595 shift register

Firmware for shift register handling

We are going to learn how to design a firmware specifically for these kinds of shift registers. This firmware is basically made for the 595 but wouldn't require a lot of modifications to be used with other integrated circuits. You'd specially have to take care about three serial pins, Latch, Clock, and Data.

Because I want to teach you each time a bit more than the exact content evoked by each chapter title, I created a very inexpensive and small random groove machine for you. Its purpose is to generate random bytes. These bytes will then be sent to the shift register in order to feed or not each LED. You'll then have then a neat random pattern of LEDs.

You can find the code for this in the `Chapter08/ Multiplexing_8Leds/` folder.

Let's check it:

```
// 595 clock pin connecting to pin 4
int CLOCK_595 = 4;

// 595 latch pin connecting to pin 3
int LATCH_595 = 3;

// 595 serial data input pin connecting to pin 2
int DATA_595 = 2;

// random groove machine variables
int counter = 0;
byte LED_states = B00000000 ;

void setup() {

  // Let's set all serial related pins as outputs
  pinMode(LATCH_595, OUTPUT);
  pinMode(CLOCK_595, OUTPUT);
  pinMode(DATA_595, OUTPUT);

  // use a seed coming from the electronic noise of the ADC
  randomSeed(analogRead(0));
}

void loop(){

  // generate a random byte
  for (int i = 0 ; i < 8 ; i++)
  {
    bitWrite(LED_states, i, random(2));
  }

  // Put latch pin to LOW (ground) while transmitting data to 595
  digitalWrite(LATCH_595, LOW);

  // Shifting Out bits i.e. using the random byte for LEDs states
  shiftOut(DATA_595, CLOCK_595, MSBFIRST, LED_states);

  // Put latch pin to HIGH (5V) & all data are pushed to outputs
  digitalWrite(LATCH_595, HIGH);
```

```
  // each 5000 loop() execution, grab a new seed for the random
function
  if (counter < 5000) counter++;
  else
  {
    randomSeed(analogRead(0));      // read a new value from analog pin
0
    counter = 0;                    // reset the counter
  }

  // make a short pause before changing LEDs states
  delay(45);
}
```

Global shift register programming pattern

First, let's check the global structure.

I first define the 3 three pins of the 595 shift register. Then, I set up each of them as output in the `setup()` block.

Then, I have a pattern that looks similar to the following:

```
digitalWrite(latch-pin, LOW)
shiftOut(data-pin, clock-pin, MSBFIRST, my_states)
digitalWrite(latch-pin, HIGH)
```

This is the usual pattern for shift- registering operations.
 The `latch-pin`, as evoked explained before, is the one providing us a way to inform the integrated circuit about the fact that we want to load it with data, and then we want it to apply these this data to its outputs.

This is a bit like saying:

- Latch-pin LOW = "Hi there, let's store what I'm about to send to you."
- Latch-pin HIGH = "Ok, now use the data I just sent to commute to your outputs or not."

Then, we have this `shiftOut()`. This function provides an easy way to send data per entire bytes packets to a specific pin (the data pin) using a specific clock/ rate speed over a particular pin (the clock pin), and given an order of transmission (`MSBFIRST` or `LSBFIRST`).

Even though we aren't going to describe the things under- the- hood here, you have to understand the MSB and LSB concept.

Let's consider a byte: 1 0 1 0 0 1 1 0.

The **MSB** is the abbreviation of **Most Significant Bit**. This bit is at the left-most position (the one of the bit having of the greatest value). Here, its value is 1.

The **LSB** is stands for the **Least Significant Bit**. This bit is at the right-most position (the bit of the smallest value) It is the bit the most at the right (the one of the bit having the smallest value). Here, its value is 0.

By fixing this argument in the shiftOut() function, we are providing special information about the sense of the transmission. Indeed, we can send the previous byte by sending these bits: 1 then, 0, then 1 0 0 1 1 0 (MSBFIRST), or by sending these bits: 0 1 1 0 0 1 0 1 (LSBFIRST).

Playing with chance and random seeds

I would like to provide an example from my personal ways of programming. Here, I'm going to describe an inexpensive and small system generating random bytes. These bytes will then be sent to the the 595, and our 8 eight-LEDs array will have a very random state.

Random, in computers, isn't really random. Indeed, the random() function is a pseudo-random number generator. It can also be named a **deterministic random bit generator** (DRBG). Indeed, the sequence is (totally) determined by a small set of initial values, including the seed.

For a particular seed, a pseudo-random number generator generates the same number sequences each time the same number sequences.

But, we you can use a trick here to disturb determinism a little bit more.

Imagine that you make the seed vary sometimes. You can also introduce an external factor of randomness into your system. As we already explained before in this book, there is always some electronic noises coming going to from to the ADC even if nothing is wired to the analog inputs. You can use that external/physical noise by reading the analog input 0, for instance.

As we now well know, analog analogRead() provides a number from 0 to 1023. This is a huge resolution for our purpose here.

This is what I have put in the firmware.

I defined a counter variable and a byte. I'm first reading the value coming from the ADC for the analog pin 0 in the setup() first. Then, I'm generating generated a random byte with a for() loop and the bitWrite() function.

I'm writing each bit of the byte LED_states using numbers generated by the random(2) number function, which gives 0 or 1, randomly. Then, I'm using use the pseudo-random-generated byte into the structure previously described.

I'm redefining each 5000 loop() execution of the seed by reading the ADC for the analog pin 0.

 If you want to use random() with computers, including Arduino and embedded systems, grab some physical and external noise.

Now, let's move further.

We can use many 74HC595 shift registers for LED handling, but imagine that you need to save some more digital pins. Okay, we saw we can save a lot using shift registers. One shift registers requires three digital pins and drives eight LEDs. It means we save five pins with each shift register, considering we wire eight LEDs.

What if you need A LOT more? What if you need to save all the other pins for switches handling, for instance?

Let's daisy chain now!

Daisy chaining multiple 74HC595 shift registers

A **daisy chain** is a wiring scheme used to link multiple devices in a sequence or even a ring.

Indeed, since we already understood a bit more about how shift registers work, we could have the idea to extend this to multiple shift registers wired together, couldn't we?

I'm going to show you how to do this by using the **ShiftOutX** library by Juan Hernandez. I had very nice results with Version 1.0, and I advise you to use this one.

You can download it here: http://arduino.cc/playground/Main/ShiftOutX. You can install it by following the procedure explained in the appendice.

Linking multiple shift registers

What would each shift register need to know about?

The serial clock, the latch, and the data are the necessary points of information that have to be transmitted all along the device chain. Let's check a schematic:

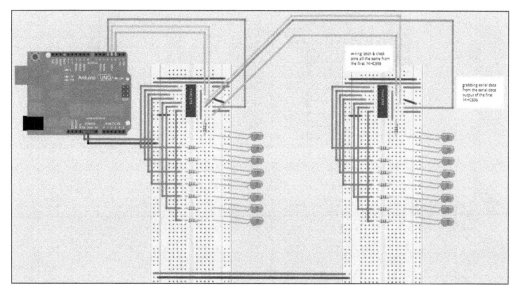

Two shift registers daisy chained driving 16 monochromatic LEDs with only three digital pins on the Arduino

I used the same colors as with the previous circuit for the clock (blue), latch (green), and serial data (orange).

The serial clock and latch are shared across the shift registers. The command/order coming from Arduino to synchronize serial communication with the clock and to tell shift registers to store or apply data received to their output has to be coherent.

The serial data coming from Arduino first goes into the first shift register, which sends the serial data to the second one. This is the core of the chaining concept.

Let's check the circuit diagram to put this in mind:

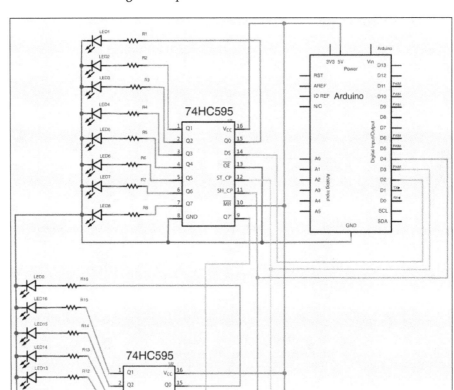

Circuit diagram of two daisy-chained shift registers driving 16 monochromatic LEDs

Firmware handling two shift registers and 16 LEDs

The firmware includes the ShiftOutX library ShiftOutX as wrote before. It provides very easy and smooth handling for daisy chaining of shift registers.

Here is the the code for the firmware.

You can find it in the Chapter08/Multiplexing_WithDaisyChain/ folder:

```
#include <ShiftOutX.h>
#include <ShiftPinNo.h>
```

```
int CLOCK_595 = 4;     // first 595 clock pin connecting to pin 4
int LATCH_595 = 3;     // first 595 latch pin connecting to pin 3
int DATA_595 = 2;      // first 595 serial data input pin connecting to
pin 2

int SR_Number = 2;     // number of shift registers in the chain

// instantiate and enabling the shiftOutX library with our circuit
parameters
shiftOutX regGroupOne(LATCH_595, DATA_595, CLOCK_595, MSBFIRST, SR_
Number);

// random groove machine variables
int counter = 0;
byte LED0to7_states = B00000000 ;
byte LED8to15_states = B00000000 ;

void setup() {

  // NO MORE setup for each digital pin of the Arduino
  // EVERYTHING is made by the library :-)

  // use a seed coming from the electronic noise of the ADC
  randomSeed(analogRead(0));
}

void loop(){

  // generate a 2 random bytes
  for (int i = 0 ; i < 8 ; i++)
  {
    bitWrite(LED0to7_states, i, random(2));
    bitWrite(LED8to15_states, i, random(2));
  }

  unsigned long int data; // declaring the data container as a very
local variable
  data = LED0to7_states | (LED8to15_states << 8); // aggregating the 2
random bytes
  shiftOut_16(DATA_595, CLOCK_595, MSBFIRST, data);  // pushing the
whole data to SRs

  // each 5000 loop() execution, grab a new seed for the random
function
```

```
    if (counter < 5000) counter++;
    else
    {
      randomSeed(analogRead(0));      // read a new value from analog pin
  0

      counter = 0;                    // reset the counter
    }

    // make a short pause before changing LEDs states
    delay(45);
}
```

The ShiftOutX library can be used in many ways. We are using it here following in the same way that we did with `shiftOut`, the library part of the core and suited for the use of only one shift register.

First, we have to include the library by using **Sketch | Import Library | ShiftOutX**.

It includes two header files at the beginning, namely, `ShiftOutX.h` and `ShiftPinNo.h`.

Then, we define a new variable storing the number of shift registers in the chain.

At last, we instantiate the ShiftOutX library by using the following code:

```
shiftOutX regGroupOne(LATCH_595, DATA_595, CLOCK_595, MSBFIRST, SR_
Number);
```

The code in `setup()` changed a bit. Indeed, there are no more setup statements for digital pins. This part is handled by the library, which can look weird but is very usual. Indeed, when you instantiated the library before, you passed three pins of Arduino as arguments, and in fact, this statement also sets up the pins as outputs.

The `loop()` block is almost the same as before. Indeed, I included again the small random groove machine with the analog read trick. But I'm creating two random bytes, this time. Indeed, this is because I need 16 values and I want to use the `shiftOut_16` function to send all my data in the same statement. It is quite easy and usual to generate bytes, then aggregate them into an `unsigned short int` datatype by using bitwise operators.

Let's detail this operation a bit.

When we generate random bytes, we have two series of 8 eight bits. Let's take the following example:

```
0 1 1 1 0 1 0 0
1 1 0 1 0 0 0 1
```

If we want to store them in one place, what could we do? We can shift one and then add the shifted one to the other one, couldn't we?

```
0 1 1 1 0 1 0 0 << 8 = 0 1 1 1 0 1 0 0 0 0 0 0 0 0 0 0
```

Then, if we add a byte using the bitwise operator (|), we get:

```
0 1 1 1 0 1 0 0 0 0 0 0 0 0 0 0
      |                            1 1 0 1 0 0 0 1
      =     0 1 1 1 0 1 0 0 1 1 0 1 0 0 0 1
```

The result seems to be a concatenation of all the bits.

This is what we are doing in this part of the code. Then we use shiftOut_16() to send all the data to the two shift registers. Hey, what should we do with the four shift registers? The same thing in the same way!

Probably we would have to shift more using << 32, << 16, and again <<8, in order to store all our the bytes into a variable that we could send using shiftOut_32() functions.

By using this library, you can have two groups, each one containing eight shift registers.

What does that mean?

It means that you can drive 2 x 8 x 8 = 128 outputs using only four pins (two latches but common serial clock and data). It sounds crazy, doesn't it?

In real life, it is totally possible to use only one Arduino to make this kind of architecture, but we would have to take care of something very important, the current amount. In this particular case of 128 LEDs, we should imagine the worst case when all the LEDs would be switched on. The amount of current driven could even burn the Arduino board, which would protect itself by resetting, sometimes. But personally, I wouldn't even try.

Current short considerations

The Arduino board, using USB power supply, cannot drive more than 500 mA. All combined pins cannot drive more than 200 mA, and no pin can drive more than 40 mA. It can vary a bit from one board type to another, but these are real, absolute maximum ratings.

We didn't make these considerations and the following calculations because, in our examples, we only used a few devices and components, but you could sometimes be tempted to build a huge device such as I made sometimes, for example, with the Protodeck controller.

Let's take an example in order to look closer at some current calculations.

Imagine that you have an LED that needs around 10 mA to bright light up correctly (without burning at the second blink!!)

This would mean you'd have 8 x 10 mA for one eight -LEDs array, driven by one 595 shift register, if all LEDs were to be switched on at the same time.

80 mA would be the global current driven by one 595 shift register from the Arduino Vcc source.

If you had more 595 shift registers, the magnitude of the current would increase. You have to know that all integrated circuits also consume current. Their consumption isn't generally taken into consideration, because it is very small. For instance, the 595 shift register circuit only consumes around 80 micro Amperes itself, which means 0.008 mA. Compared to our LEDs, it is negligible. Resistors consume current too, even if they are often used to protect LEDs, they are very useful.

Anyway, we are about to learn another very neat and smart trick that can be used for monochromatic or RGB LEDs.

Let's move to a world full of colors.

Using RGB LEDs

RGB stands for Red, Green, and Blue, as you were probably guessing.

I don't talk about LEDs that can change their color according to the voltage you apply to them. LEDs of this kind exists, but as far as I experimented, these aren't the way to go, especially while still learning steps.

I'm talking about common cathode and common anode RGB LEDs.

Some control concepts

What do you need to control an LED?

You need to be able to apply a current to its legs. More precisely, you need to be able to create a difference of potential between its legs.

The direct application of this principle is what we have already tested in the first part of this chapter, which remind us how we can switch on an LED: we you need to control the current using digital output pins of our Arduino, knowing the LED we want to control has its node wired to the output pin and its cathode wired to the ground, with a resistor on the line too.

We can discuss the different ways of controls, and you are going to understand that very quickly with the next image.

In order to make a digital output sourcing current, we need to write with `digitalWrite` to it a value of `HIGH`. In that this case, the considered digital output will be internally connected to a 5 V battery and will produce a voltage of 5 V. That means that the wired LED between it and the ground will be fed by a current.

In the other case, if we apply 5 V to an LED and if we want to switch it on, we need to write a value of `LOW` to the digital pin to which it is linked. In this case, the digital pin will be internally connected to the ground and will sink the current.

These are the two ways of controlling the current.

Check the following diagram:

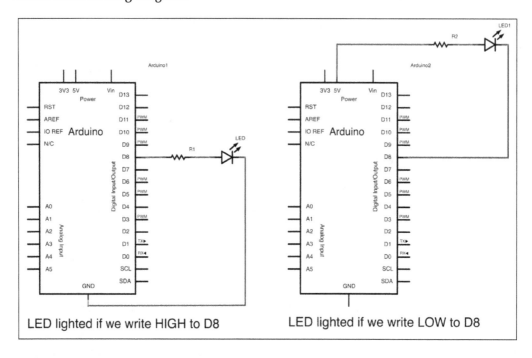

LED lighted if we write HIGH to D8

LED lighted if we write LOW to D8

Different types of RGB LEDs

Let's check the two common RGB LEDs:

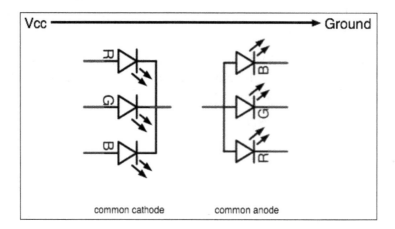

common cathode   common anode

There are basically three LEDs in one package, with different types of wiring inside. The way of making this package isn't really about wiring inside, but I won't debate that here.

If you followed me correctly, you may have guessed that we need more digital outputs to connect RGB LEDs. Indeed, the previous section talked about saving digital pins. I guess you understand why it could be important to save pins and to plan our circuit architectures carefully.

Lighting an RGB LED

Check this basic circuit:

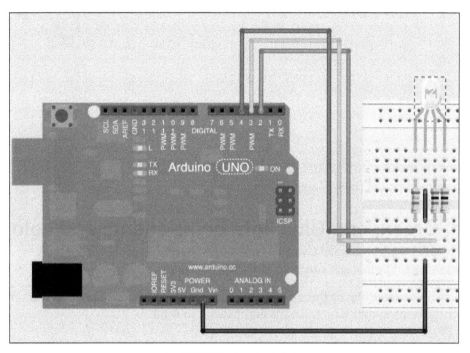

An RGB LED wired to Arduino

Check the code now. You can find it in the `Chapter08/One_RGB_LED/` folder.

```
int pinR = 4; // pin related to Red of RGB LED
int pinG = 3; // pin related to Green of RGB LED
int pinB = 2; // pin related to Blue of RGB LED

void setup() {

  pinMode(pinR, OUTPUT);
  pinMode(pinG, OUTPUT);
  pinMode(pinB, OUTPUT);
}
```

```
void loop() {

  for (int r = 0 ; r < 2 ; r++)
  {
    for (int g = 0 ; g < 2 ; g++)
    {
      for (int b = 0 ; b < 2 ; b++)
      {
        digitalWrite(pinR,r); // turning red pin to value r
        digitalWrite(pinG,g); // turning green pin to value g
        digitalWrite(pinB,b); // turning blue pin to value b

        delay(150); // pausing a bit
      }
    }
  }

}
```

Again, some tips are present inside this code.

Red, Green, and Blue light components and colors

First, what is the point here? I want to make the RGB LED cycle through all the possible states. Some math can help to list all the states.

We have an ordered list of three elements, each one of which can be on or off. Thus, there are 23 states, that which means eight states in total:

R	G	B	Resulting color
Off	Off	Off	OFF
Off	Off	On	Blue
Off	On	Off	Green
Off	On	On	Cyan
On	Off	Off	Red
On	Off	On	Purple
On	On	Off	Orange
On	On	On	White

Only by switching each color component on or off, can we change the global RGB LED state.

Don't forget that the system works exactly as if we were controlling three monochromatic LEDS through three digital outputs from Arduino.

First, we define three variables storing the different colors LED connectors.

Then, in the `setup()`, we set those 3 three pins as output.

Multiple imbricated for() loops

At last, the `loop()` block contains triple-imbricated `for()` loops. What's that? It is nice efficient way to be sure to match all the cases possible. It is also an easy way to cycle each number possible. Let's check the first steps, in order to understand this imbricated loops concept better.:

- 1st step: **r = 0, g = 0, and b = 0** implies everything is OFF, then pauses for 150ms in that state
- 2nd step: **r = 0, g = 0, and b = 1** implies only BLUE is switched on, then pauses for 150ms in that state
- 3rd step: **r = 0, g = 1, and b = 0** implies only GREEN is switched on, then pauses for 150ms in that state

The innermost loop is always the one executed the most number of times.

Is that okay? You bet, it is!

You also may have noticed that I didn't write HIGH or LOW as arguments for the `digitalWrite()` function. Indeed, HIGH and LOW are constants defined in the Arduino core library and are only replace the values 1 and 0, respectively.

In order to prove this, and especially to show you for the first time where the Arduino core files sit, the important file to check here is `Arduino.h`.

On a Windows systems, it can be found in the `Arduino` folder inside some subdirectories, depending upon the version of the IDE.

On OS X, it is in `Arduino.app/Contents/Resources/Java/hardware/arduino/cores/arduino/Arduino.h`. We can see the content of an application package by right-clicking on the package itself.

In this file, we can read a big list of constants, among many other definitions.

And finally, we can retrieve the following:

```
#define HIGH 0x1
#define LOW  0x0
```

Yes, the HIGH and LOW keywords are just constants for 1 and 0.

This is the reason why I'm directly feeding digitalWrite() with 0 and 1 through the imbricated loops, cycling over all the states possible for each LED, and as a consequence, over all states for the RGB LED.

Using this concept, we are going to dig further by making an LED array.

Building LED arrays

LED arrays are basically LEDs wired as a matrix.

We are going to build a 3 x 3 LEDs matrix together. This is not that hard, and we'll approach this task with a really nice, neat and smart concept that can really optimize your hardware designs.

Let's check the simplest schematic of this book:

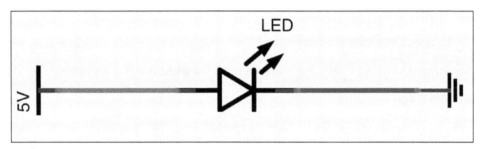

An LED can blink when a current feeds it, when a voltage is applied to its legs

In order to switch off the LED shown in the preceding screenshot, we can stop to create the 5 V current at its node. No voltage means no current feeding. We can also cut the circuit itself to switch off the LED. And at last, we can change the ground by putting adding a 5 V source current.

This means that as soon as the difference of potential is cancelled, the LED is switched off.

An LED array is based on these double controls possible.

We are going to introduce a new component right here, the transistor.

A new friend named transistor

A **transistor** is a special component that we introduced a bit in the first part of this book.

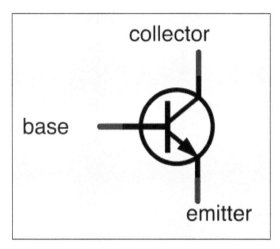

The usual NPN transistor with its three legs

This component is usually used in three main cases:

- As a digital switch in a logical circuit
- As a signal amplifier
- As a voltage stabilizer combined with other components

Transistors are the most widespread components in the world. They are not only used as discrete components (independent ones) but are also combined with many others into a high-density system, for instance, in processors.

The Darlington transistors array, ULN2003

We are going to use a transistor here, as included inside an integrated circuit named ULN2003. What a pretty name! A more explicit one is **High-current Darlington Transistors Array**. Ok, I know that doesn't help!

```
          ULN2003A
    1                        16
   ─── In 1        Out 1 ───
    2                        15
   ─── In 2        Out 2 ───
    3                        14
   ─── In 3        Out 3 ───
    4                        13
   ─── In 4        Out 4 ───
    5                        12
   ─── In 5        Out 5 ───
    6                        11
   ─── In 6        Out 6 ───
    7                        10
   ─── In 7        Out 7 ───
    8                         9
   ─── 0V           COM ───
```

Its datasheet can be found at

http://www.ti.com/lit/ds/symlink/uln2003a.pdf.

It contains seven pins named inputs and seven named outputs. We can see also a 0 V pin (the number 8) and the COM pin 9 too.

The principle is simple and amazing:

- 0 V has to be connected to the ground
- If you apply 5 V to the input n, the output n is commuted to ground

If you apply 0 V to the input n, the output n will get disconnected.

This can easily be used as a current sink array of switches.

Combined with 74HC595, we'll drive our 3 x 3 LED matrix right now:

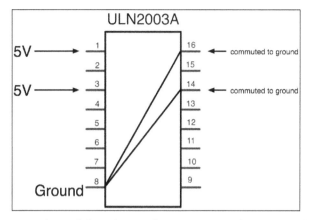

A case where inputs 1 and 2 are fed, resulting in the commutation of outputs 1 and 2 (pin 16 and 14)

The LED matrix

Let's check how we can wire our matrix, keeping in mind that we have to be able to control each LED independently, of course.

This kind of design is very usual. You can easily find ready made matrices of LEDs wired like this, sold in packages with connectors available related to rows and columns.

An LED matrix is basically an array where:

- Each row pops out a connector related to all the anodes of that row
- Each column pops out a connector related to all the cathodes of that column

This is not law, and I found some matrices wired totally in the opposite way and sometimes quite strangely. So, be careful and check the datasheet. Here, we are going to study a very basic LED matrix in order to dig into that concept:

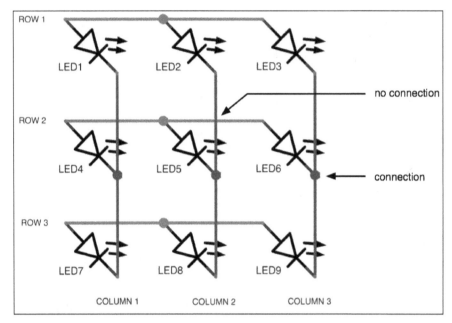

A basic 3 x 3 LED matrix

Let's look at the LED matrix architecture concept.

How can we control it? By controlling, I mean addressing the good LED to a good behavior, from being switched on or off.

Let's imagine that, if we want to light up the **LED 2**, we have to:

- Connect **ROW 1** to 5 V
- Connect **COLUMN 2** to the ground

Good! We can light up that **LED 2**.

Let's move further. Let's imagine that, if we want to light up the **LED 2** and **LED 4**, we have to:

- Connect **ROW 1** to 5 V
- Connect **COLUMN 2** to the ground
- Connect **ROW 2** to 5 V
- Connect **COLUMN 1** to the ground

Did you notice something?

If you follow the steps carefully, you should have something strange on your matrix:

LED 1, LED 2, LED 4, and **LED5** would be switched ON

Problem appeared: if we put 5 V to the **ROW 1**, how can you distinguish **COLUMN 1** and **COLUMN 2**?

We are going to see that it isn't hard at all and that it just uses a small trick related to our persistence of vision.

Cycling and POV

We can take care of the problem encountered in the previous section by cycling our matrix quickly.

The trick is switching ON only one column at a time. This could also work by switching ON only one row at a time, of course.

Let's take our previous problem: If we want to light up the **LED 2** and **LED 4**, we have to:

- Connect **ROW 1** to 5 V and **COLUMN 1** to 5 V only
- Then, put connect **ROW 2** to 5 V and **COLUMN 2** to 5 V only

If we we are doing that this very quickly, our eyes won't see that there is only one LED switched on at a time.

The pseudo code would be:

For each column

Switch On the column

For each row

Switch on the row if the corresponding LED has to be switched On

The circuit

First, the circuit has to be designed. Here is how it looks:

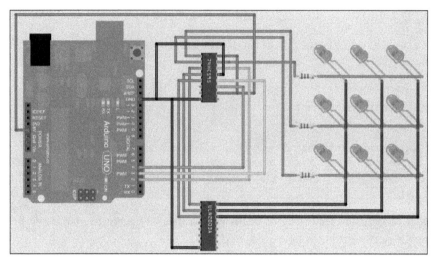

Arduino wired to a 595 shift register driving each row and column through an ULN2003

Let's now check the circuit diagram:

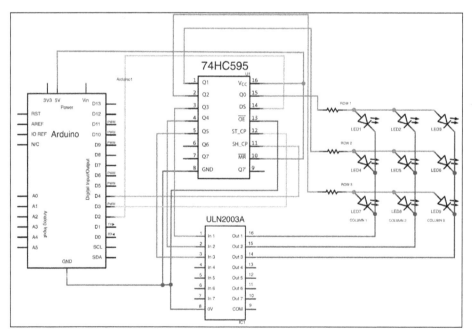

Circuit diagram showing the handling of matrix rows and columns

We have the now well-known shift register 74HC595.

This one is wired to a ULN2003 shift register and to the matrix' rows, the ULN2003 being wired to the columns of the matrix.

What is that design pattern?

The shift register grabs data from the serial protocol-based messages sent by the Arduino from its digital pin 2. As we tested before, the shift register is clocked to Arduino, and as soon as its latch pin is connected to HIGH (=(equal to 5 V), it drives an output to 5V or not, depending upon the data sent to it by Arduino. As a consequence, we can control each row of the matrix, feeding them rows with 5V or not through the data sent to the shift register.

In order to switch on LEDs, we have to close the circuit on which they are plugged, the electrical line, I mean. We can feed the **ROW 1** with a 5V current, but if we don't put this or that column to the ground, the circuit won't be closed and no LED will be switched on. Right?

The ULN2003 was made precisely for the purpose of ground commutation, as we already saw. And if we feed 5V to one of its input, it commutes the corresponding out n pin to the ground. So, with our 595 shift registers, we can control the 5V commutation for rows, and the ground commutation for columns.
We now have total control over our matrix.

Especially, we are going to check the code, including the power cycle of columns previously explained.

The 3 x 3 LED matrix code

You can find the following 3 x 3 LED matrix code in the `Chapter08/ LedMatrix3x3/` folder:

```
int CLOCK_595 = 4;    // first 595 clock pin connecting to pin 4
int LATCH_595 = 3;    // first 595 latch pin connecting to pin 3
int DATA_595 = 2;     // first 595 serial data pin connecting to pin 2

// random groove machine variables
int counter = 0;
boolean LED_states[9] ;


void setup() {

  pinMode(LATCH_595, OUTPUT);
```

```
    pinMode(CLOCK_595, OUTPUT);
    pinMode(DATA_595, OUTPUT);

    // use a seed coming from the electronic noise of the ADC
    randomSeed(analogRead(0));
}

void loop() {

    // generate random state for each 9 LEDs
    for (int i = 0 ; i < 9 ; i++)
    {
      LED_states[i] = random(2) ;
    }

    // initialize data at each loop()
    byte data = 0;
    byte dataRow = 0;
    byte dataColumn = 0;
    int currentLed = 0;


    // cycling columns
    for (int c = 0 ; c < 3 ; c++)
    {
      // write the 1 at the correct bit place (= current column)
      dataColumn = 1 << (4 - c);

      // cycling rows
      for (int r = 0 ; r < 3 ; r++)
      {
        // IF that LED has to be up, according to LED_states array
        // write the 1 at the correct bit place (= current row)
        if (LED_states[currentLed]) dataRow = 1 << (4 - c);

        // sum the two half-bytes results in the data to be sent
        data = dataRow | dataColumn;

        // Put latch pin to LOW (ground) while transmitting data to 595
        digitalWrite(LATCH_595, LOW);

        // Shifting Out bits
        shiftOut(DATA_595, CLOCK_595, MSBFIRST, data);
```

```
      // Put latch pin to HIGH (5V) & all data are pushed to outputs
      digitalWrite(LATCH_595, HIGH);


      dataRow = 0; // resetting row bits for next turn
      currentLed++;// incrementing to next LED to process
    }

    dataColumn = 0;// resetting column bits for next turn
  }


  // each 5000 loop() execution, grab a new seed for the random
function
  if (counter < 5000) counter++;
  else
  {
    randomSeed(analogRead(0));     // read a new value from analog pin
0
    counter = 0;                   // reset the counter
  }

  // pause a bit to provide a cuter fx
  delay(150);
}
```

This code is quite self-explanatory with comments, but let's check it out a bit more.

The global structure reminds the one in Multiplexing_8Leds.

We have an integers array named LED_states. We are storing data for each LED state inside of it. The setup() block is quite easy, defining each digital pin used in the communication with the 595 shift- register and then grabbing a random seed from the ADC. The loop() is a bit more tricky. At first, we generating nine random values and store them in the LED_states array. Then, we initialize/define some values:

- data is the byte sent to the shift register
- dataRow is the part of the byte handling row state (commuted to 5V or not)
- dataColumn is the part of the byte handling column state (commuted to the ground or not)
- currentLed keeps the trace of the current handled handled by the LED

Then, those imbricated loops occur.

For each column (first for() loop), we activate it the loop by using a small/cheap and fast bitwise operator:

```
dataColumn = 1 << (4 - c);
```

`(4 - c)` goes from 4 to 2, all along this first `loop()`; function; then, `dataColumn` goes from: 0 0 0 1 0 0 0 0 to 0 0 0 0 1 0 0 0, and at last 0 0 0 0 0 1 0 0.

What's going on right here? It is all about coding.

The first three bits (beginning at the left, the MSB bit) handle the rows of our matrix. Indeed the three rows are connected to the Q0, Q1, and Q2 pins of the 595 shift register.

The second three-bit group handles the ULN2003, which itself handles the columns.

By feeding 5 V from Q0, Q1, and Q2 of the 595, we handle rows. By feeding 5 V from Q3, Q4, and Q5 of the 595, we handle columns through the ULN2003.

Good!

We still have two bits not unused bits right here, the last two.

Let's take look at our our code again.

At each column turn of the for() loop, we move the bit corresponding to the column to the right, commuting each column to the ground cyclically.

Then, for each column, we cycle the row on the same mode, testing the state of the corresponding LED that we have to push to the 595. If the LED has to be switched on, we store the corresponding bit in the dataRow variable with the same bitwise operation trick.

Then, we sum those two parts, resulting in the data variable.

For instance, if we are on the second row and the second column and the LED has to be switched on, then the data stored will be:

0 1 0 0 1 0 0 0.

If we are at (1,3), then the data stored will be data will store:

1 0 0 0 0 1 0 0.

Then, we have the pattern that adds Latch to LOW, shifting out bits stored in data to the shift- register, and then putting adds Latch to HIGH to commit data to the Q0 to Q7 outputs, feeding the right elements in the circuits.

At the end of each row handled, we reset the three bits corresponding to the first three rows and increment the `currentLed` variable.

At the end of each column handled, we reset the three bits corresponding to the next three columns.

This global imbricated structure makes us ensures that we'll have only one LED switched on at a time.

What is the consequence of the current consumption?

We'll only have one LED fed, which means we'll have our maximum consumption potentially divided by nine. Yes, that sounds great!

Then, we have the pattern grabbing that grabs a new seed, each 5000 loop() turn.

We just learned how to handle LED matrices quite easily and to reduce our power consumption at the same time.

But, I'm not satisfied. Usually, creators and artists are generally never completely satisfied, but here, trust me it's different; we could do better things than just switching on and off LEDs. We could dim them too and switch them from a very low intensity to a very high one, making some different shades of light.

Simulating analog outputs with PWM

As we know very well by now, it's okay to switch on/off LEDs, and as we are going to see in the next chapter, to switch on/off many things too by using digital pins as output on the Arduino.

We also know how to read states from digital pins set up as inputs, and even values from 0 to 1023 from the analog inputs from in the ADC.

As far as we know, there isn't analog output on the Arduino.

What would an analog output add? It would provide a way to write values other than only 0 and 1, I mean 0 V and 5 V. This would be nice but would require an expensive DAC.

Indeed, there isn't a DAC on Arduino boards.

The pulse-width modulation concept

The **pulse-width modulation** is a very common technique used to mimic analog output behavior.

Let's put that another way.

Our digital outputs can only be at 0 V or 5 V. But at a particular time-interval, if we switch them on/off quickly, then we can calculate a mean value depending on the time passed at 0 V or 5 V. This mean can easily be used as a value.

Check the following schematic to know know more about the concept of duty cycle:

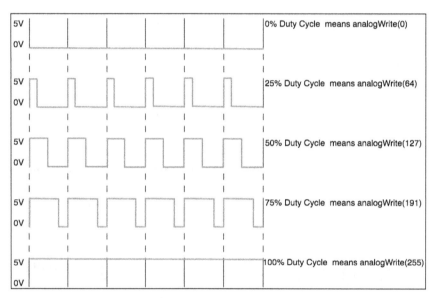

The concept of duty cycle and PWM

The mean of the time spent at 5V defines the duty cycle. This value is the mean time when the pin is at 5V and is given as a percentage.

`analogWrite()` is a special function that can generate a steady square wave at a specific duty cycle until the next call.

According to the Arduino core documentation, the PWM signal pulses at a frequency of 490 Hz. I didn't (yet) verify this, but it would really only be possible with an oscilloscope, for instance.

Be careful: PWM isn't available on every pin of your board!

For instance, Arduino Uno and Leonardo provide PWM on digital pins numbers 3, 5, 6, 9, 10, and 11.

You have to know this before trying anything.

Dimming an LED

Let's check a basic circuit in order to test PWM:

Let's look at the circuit diagram, even if it's obvious:

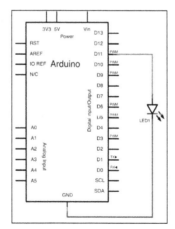

We'll use the Fading example by David A. Mellis and modified by Tom Igoe. Check it in **File | examples | 03.Analog | Fading**. We are going to change the `ledPin` value from 9 to 11 to fit our circuit.

Here it is, modified:

```
int ledPin = 11;     // LED connected to digital pin 11 (!!)

void setup()  {
  // nothing happens in setup
}

void loop()  {
  // fade in from min to max in increments of 5 points:
  for(int fadeValue = 0 ; fadeValue <= 255; fadeValue +=5) {
    // sets the value (range from 0 to 255):
    analogWrite(ledPin, fadeValue);
    // wait for 30 milliseconds to see the dimming effect
    delay(30);
  }

  // fade out from max to min in increments of 5 points:
  for(int fadeValue = 255 ; fadeValue >= 0; fadeValue -=5) {
    // sets the value (range from 0 to 255):
    analogWrite(ledPin, fadeValue);
    // wait for 30 milliseconds to see the dimming effect
    delay(30);
  }
}
```

Upload it, test it, and love it!

A higher resolution PWM driver component

Of course, there are components providing higher resolutions of PWM. Here, with native Arduino boards, we have an 8-bit resolution (256 values). I wanted to point out to you the Texas Instrument, TLC5940. You can find its datasheet here: `http://www.ti.com/lit/ds/symlink/tlc5940.pdf`.

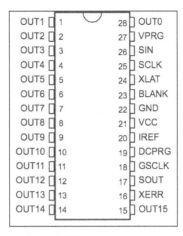

TLC5950, the 16-channel LED driver that provides PWM control

Be careful, it is a constant-current sink driver. This means that it sinks the current and does not feed the current. For instance, you'd have to connect cathodes of your LEDs to the OUT0 and OUT15 pins, not anodes. If you want to use a specific driver like that, you won't use analogWrite(), of course. Why? Because this driver works as a shift register, wired through a serial connection with our Arduino.

I'd suggest using a nice library named tlc5940arduino, and available on Google code at

http://code.google.com/p/tlc5940arduino/

We'll see, in the third part of this book, how to write messages on LED matrices. But, there is also a nice way to use highest resolution displays: LCD.

Quick introduction to LCD

LCD means **Liquid Crystal Display**. We use LCD technology everyday in watches, digicode display, and so on. Look around you, and check these small or great LCDs.

There exist two big families of LCD displays:

- Character LCD is based on a matrix of characters (columns x rows)
- Graphical LCD , is based on a pixel matrix

We can find a lot of printed circuit boards that include an LCD and the connectors to interface them with Arduino and other systems for cheap, nowadays.

There is now a library included in the Arduino Core that is really easy to use. Its name is **LiquidCrystal**, and it works with all LCD displays that are compatible with the Hitachi HD44780 driver. This driver is really common.

Hitachi developed it as a very dedicated driver, that includes a micro-controller itself, specifically to drive alphanumeric characters LCDs and to connect to the external world easily too, which can be done by a specific link using, usually, 16 connectors, including power supply for the external circuit itself and the backlight supply too:

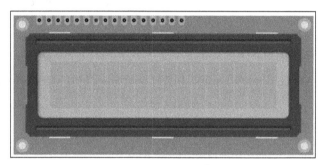

A 16 x 2 character LCD

We are going to wire it and display some messages on it.

HD44780-compatible LCD display circuit

Here is the basic circuit of the HD44780-compatible LCD display:

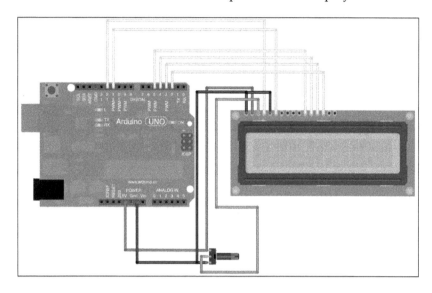

A 16 x 2 character LCD wired to Arduino and a potentiometer controlling its contrast

The corresponding circuit diagram is as follows:

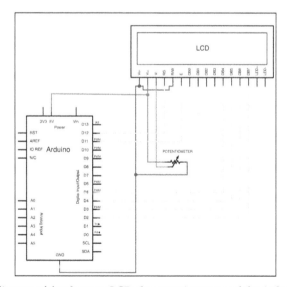

Circuit diagram of the character LCD, the potentiometer, and the Arduino board

LED+ and LED- aren't necessary as far as you have sufficient light. Using the potentiometer, you can also set the contrast of the LCD in order to have enough readability.

By the way, LED+ and LED- are, respectively, backlight anode and backlight cathode of the internal LED used for the backlight. You can drive these from Arduino, but it can lead to more consumption. Please read the LCD instructions and datasheet carefully.

Displaying some random messages

Here is some neat firmware. You can find it in the Chapter08/basicLCD/ folder:

```
#include <LiquidCrystal.h>

String manyMessages[4];
int counter = 0;

// Initialize the library with pins number of the circuit
// 4-bit mode here without RW
LiquidCrystal lcd(12, 11, 5, 4, 3, 2);

void setup() {

  // set up the number of column and row of the LCD
  lcd.begin(16, 2);

  manyMessages[0] = "I am the Arduino";
  manyMessages[1] = "I can talk";
  manyMessages[2] = "I can feel";
  manyMessages[3] = "I can react";

  // shaking the dice!
  randomSeed(analogRead(0);
}

void loop() {

  // set the cursor to column 0 and row 0
  lcd.setCursor(0, 0);

  // each 5s
```

```
if (millis() - counter > 5000)
{
  lcd.clear(); // clear the whole LCD
  lcd.print(manyMessages[random(4)]); // display a random message
  counter = millis();  // store the current time
}

// set the cursor to column 0 and row 1
lcd.setCursor(0, 1);
// print the value of millis() at each loop() execution
lcd.print("up since: " + millis() + "ms");
}
```

First, we have to include the `LiquidCrystal` library. Then, we define two variables:

- `manyMessages` is an array of String for message storage
- `counter` is a variable used for time tracing

Then, we initialize the `LiquidCrystal` library by passing some variables to its constructor, corresponding to each pin used to wired the LCD to the Arduino. Of course, the order of pins matters. It is: `rs`, `enable`, `d4`, `d5`, `d6`, and `d7`.

In the `setup()`, we define the size of the LCD according to the hardware, here that would be 16 columns and two rows.

Then, we statically store some messages in each element of the String array.

In the `loop()` block, we first place the cursor to the first place of the LCD.

We test the expression `(millis() - counter > 5000)`, and if it is true, we clear the whole LCD. Then, I'm printing a message defined by chance. Indeed, `random(4)` produces a pseudo-random number between 0 and 3 , and that index being random, we print a random message to the LCD from among the four defined in `setup()` to the LCD, on the first row.

Then, we store the current time in order to be able to measure the time passed since the last random message was displayed.

Then, we put the cursor at the first column of the second row, then, we print a String composed by constant and variable parts displaying the time in milliseconds since the last reset of the Arduino board.

Summary

In this long chapter, we learned to deal with many things, including monochromatic LEDs to RGB LEDs, using shift registers and transistor arrays, and even introduce the LCD display. We dug a bit deeper into displaying visual feedbacks from the Arduino without necessarily using a computer.

In many cases of real life design, we can find projects using Arduino boards totally standalone and, without a computer. Using special libraries and specific components, we now know that we can make our Arduino feeling, expressing, and reacting.

In the following chapter, we are going to explain and dig into some other concepts, such as making Arduino move and eventually generating sounds too.

9
Making Things Move and Creating Sounds

If the Arduino board can listen and feel with sensors, it can also react by making things move.

By the movement concept, I mean both of the following:

- Object movements
- Air movements producing sounds

We are going to learn how we can control small motors named **servo**, and how we can deal with high-current control by using transistors.

Then we'll start talking about the basics of sound generation. This is a requirement before trying to produce any sounds, even the simplest ones. This is the part where we'll describe analog and digital concepts.

At last, we'll design a very basic random synthesizer controllable using MIDI. We'll also introduce a very nice library called **PCM** that provides a simple way to add sample playing features to your 8-bit microcontroller.

Making things vibrate

One of the simplest projects we can introduce here is the use of a small piezoelectric sensor.

This is the first basic tangible action we design here. Of course, we already designed many of the visual feedback, but this is our first real-world object that moves the firmware.

This kind of feedback can be very useful in nonvisual contexts. I designed a small project for a person who wanted to send a feedback to visitors in his reactive installation. The visitor had to put on a t-shirt that included some electronics attached, such as a LilyPad and some piezoelectric sensors. The LED feedback wasn't the solution we used before to send feedback to the wearer, and we decided to send a vibration. These piezoelectric sensors were distributed on each side of the t-shirt to produce different feedback in response to different interactions.

But wouldn't I have made a mistake talking about sensors vibrating?

The piezoelectric sensor

A piezoelectric sensor is a component that uses the piezoelectric effect.

This effect is defined as the linear electromechanical interaction between the mechanical and electrical state in some specific materials.

Basically, a mechanical action on this device generates electricity, making it usable for movement and vibration detection. But the nice thing here is that the effect is reciprocal—if you apply a current to it, it will vibrate.

This is why we are using a piezoelectric sensor here. We are using it as a vibration generator.

Piezoelectric sensors are also often used as a tone generator. We will dig deeper into the relationship between air vibrations and sound a bit later, but it is important to mention it here too.

Wiring a vibration motor

Piezoelectric sensors usually consume around 10 mA to 15 mA, which is very small.

Of course, you need to check the proper datasheet of the device you are going to use. I have had good results with the one from **Sparkfun** (https://www.sparkfun.com/products/10293). The wiring could not be simpler—there are only two legs. The following image shows how the piezoelectric sensor/vibrator is wired to Arduino via a PWM-capable digital pin:

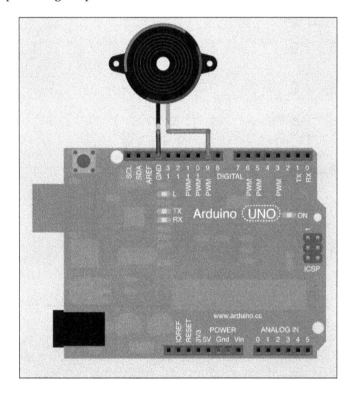

Please note that I have wired the piezoelectric device to a PWM-capable digital pin. I explained PWM in the previous chapter.

Here is the circuit schematic. This piezoelectric component is displayed as a small buzzer/speaker:

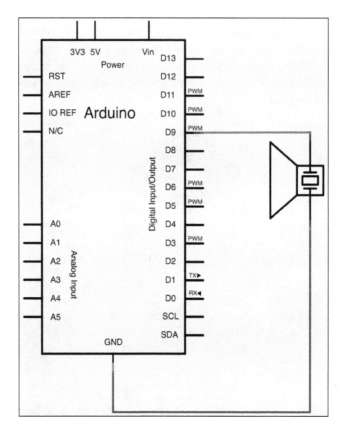

Of course, since we are going to use PWM, it means that we are going to simulate an analog output current. Considering the duty-cycle concept, we can feed the piezoelectric device using the `analogWrite()` function and then feed it with different voltages.

Firmware generating vibrations

Check the firmware. It is also available in the `Chapter09/vibrations/` folder.

```
int piezoPin = 9;
int value = 0;  // stores the current feed value
int incdec = 1; // stores the direction of the variation

void setup() {
```

```
}

void loop() {

  // test current value and change the direction if required
  if (value == 0 || value == 255) incdec *= -1;

  analogWrite(piezoPin, value + incdec);
  delay(30);
}
```

We are using the `analogWrite()` function here again. This function takes the digital pin as an argument and value. This value from 0 to 255 is the duty cycle. It basically simulates an analog output.

We use it the usual way with the `incdec` (stands for increment-decrement) parameter. We store the increment value we want to use at each `loop()` execution.

This increment changes when the value reaches its boundaries, 0 or 255, and is inverted, providing a cheap way to make a cycle from 0 to 255, then to 0, then to 255, and so on.

This firmware makes the piezoelectric device vibrate cyclically from a low rate to a higher rate.

Let's control bigger motors now.

Higher current driving and transistors

We talked about transistors in the previous chapter. We used them as digital switches. They can also be used as amplifiers, voltage stabilizers, and many other related applications.

You can find transistors almost everywhere and they are quite cheap. You can find the complete datasheet at `http://www.fairchildsemi.com/ds/BC/BC547.pdf`.

The following is a basic diagram explaining how transistors work:

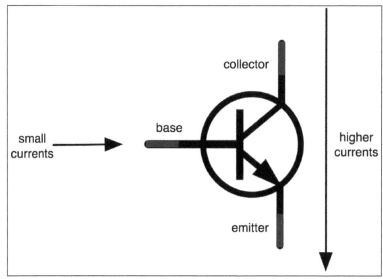

The transistor used as a digital switch in logical circuits

A transistor has the following legs:

- The collector
- The base
- The emitter

If we saturate the base by applying a 5 V power supply to it, all the current coming from the collector will be transmitted through the emitter.

When used like this, the NPN transistor is a nice way to switch on/off high current that Arduino wouldn't have been able to drive by itself. By the way, this switch is controllable with Arduino because it only requires a very small amount of current to be provided to the base of the transistor.

 Sending 5 V to the transistor base closes the circuit. Putting the transistor base to ground opens the circuit.

In any case, where you need to have an external power supply to drive motors, we use this kind of design pattern.

Let's now learn about small current servos and then move further using transistors.

Controlling a servo

A **servomotor** is also defined as a rotary actuator that allows for very fine control of angular positions.

Many servos are widely available and quite cheap. I have had nice results with a 43 R servo, by Spring Model Electronics. You can find the datasheet at `http://www.` `sparkfun.com/datasheets/Robotics/servo-360_e.pdf`.

Servos can drive a great amount of current. This means that you wouldn't be able to use more than one or two on your Arduino board without using an external source of power.

When do we need servos?

Whenever we need a way to control a position related to a rotation angle, we can use servos.

Servos can not only be used to move small parts and make objects rotate, but can also be used to move the object including them. Robots work in this fashion, and there are many Arduino-related robot projects on the Web that are very interesting.

In the case of robots, the servo device case is fixed to a part of an arm, for instance, and the other part of the arm is fixed to the rotating part of the servo.

How to control servos with Arduino

There is a nice library that should be used at first, named `Servo`.

This library supports up to 12 motors on most Arduino boards and 48 on the Arduino Mega.

By using other Arduino boards over Mega, we can figure out some software limitations. For instance, pins 9 and 10 cannot be used for PWM's `analogWrite()` method (`http://arduino.cc/en/Reference/analogWrite`).

Servos are provided in three-pin packages:

- 5 V
- Ground
- Pulse; that is, control pin

Basically, the power supply can be easily provided by an external battery, and the pulse still remains the Arduino board.

Let's check the basic wiring.

Wiring one servo

The following diagram is that of a servo wired to an Arduino for both power supply and control:

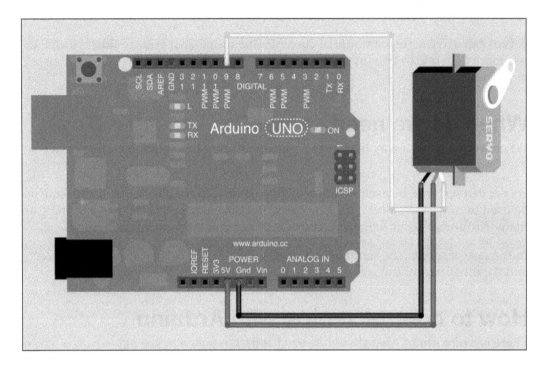

The corresponding circuit diagram is as follows:

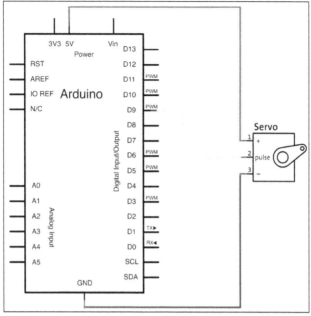

One servo and Arduino

We are basically in a very common digital output-based control pattern.

Let's check the code now.

Firmware controlling one servo using the Servo library

Here is a firmware that provides a cyclic movement from 0 degrees to 180 degrees. It is also available in the Chapter09/OneServo/ folder.

```
#include <Servo.h>

Servo myServo;   // instantiate the Servo object
int angle = 0;   // store the current angle

void setup()
{
```

```
    // pin 9 to Servo object myServo
    myServo.attach(9);
}


void loop()
{
  for(angle = 0; angle < 180; angle += 1)
  {
    myServo.write(angle);
    delay(20);
  }
  for(angle = 180; angle >= 1; angle -=1)
  {
    myServo.write(angle);
    delay(20);
  }
}
```

We first include the `Servo` library header.

Then we instantiate a `Servo` object instance named `myServo`.

In the `setup()` block, we have to make something special. We attach pin 9 to the `myServo` object. This explicitly defines the pin as the control pin for the `Servo` instance `myServo`.

In the `loop()` block, we have two `for()` loops, and it looks like the previous example with the piezoelectric device. We define a cycle, progressively incrementing the angle variable from 0 to 180 and then decrementing it from 180 to 0, and each time we pause for 20 ms.

There is also a function not used here that I want to mention, `Servo.read()`.

This function reads the current angle of the servo (that is, the value passed to the last call to `write()`). This can be useful if we are making some dynamic stuff without storing it at each turn.

Multiple servos with external power supply

Let's imagine we need three servos. As explained before, servos are motors, and motors convert current into movement, driving more current than other kinds of devices such as LEDs or sensors.

If your Arduino project requires a computer, you can supply power to it with the USB as long as you don't go beyond the 500 mA limit. Beyond this, you'd need to use an external power supply for some or all parts of your circuit.

Let's see how it goes with three servos.

Three servos and an external power supply

An external power supply can be batteries or a wall adapter power supply.

We are going to use basic AA batteries here. This is also a way to supply Arduino if you don't need a computer and want Arduino to be autonomous. We will consider this option in the third part of this book about more advanced concepts.

Let's check the wiring for now:

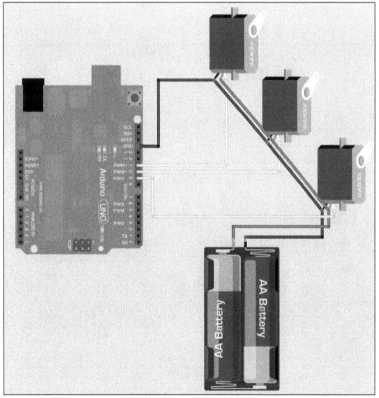

Three servos wired to an Arduino, and power supplied by two AA batteries

In cases like this, we have to wire the grounds together. Of course, there is only one current source supply for the servos — the two AA batteries.

Let's check the circuit diagram:

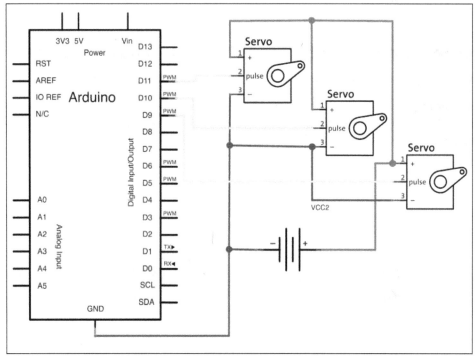

Three servos, two AA batteries, and an Arduino

Driving three servos with firmware

Here is an example of firmware for driving three servos:

```
#include <Servo.h>

Servo servo01;
Servo servo02;
Servo servo03;

int angle;

void setup()
{
```

```
    servo01.attach(9);
    servo02.attach(10);
    servo03.attach(11);
}

void loop()
{
    for(angle = 0; angle < 180; angle += 1)
    {
        servo01.write(angle);
        servo02.write(135-angle/2);
        servo03.write(180-angle);

        delay(15);
    }
}
```

This very minimal firmware is also available in the `Chapter09/Servos/` folder.

We first instantiate our three servos and attach one pin for each in the `setup()` block.

In `loop()`, we play with angles. As a new approach for generative creation, I defined one variable only for the angle. This variable cyclically goes from 0 to 180 in each `loop()` turn.

The servo attached to pin 9 is driven with the angle value itself.

The servo attached to pin 10 is driven with the value [135-(angle/2)], varying itself from 135 to 45.

Then, the servo attached to pin 11 is driven with the value [180-angle], which is the opposite movement of the servo attached to pin 9.

This is also an example to show you how we can easily control one variable only, and program variations around this variable each time; here, we are making angles vary and we are combining the angle variable in different expressions.

Of course, we could control the servo position by using an external parameter, such as a potentiometer position or distance measured. This will combine concepts taught here with those in *Chapter 5, Sensing with Digital Inputs*, and *Chapter 6, Sensing the World–Feeling with Analog Inputs*.

Let's learn a bit more about step motors.

Controlling stepper motors

Stepper motor is the common name for a **step motor**. They are motors that are controllable using small steps.

The full rotation is divided into a number of equal steps and the motors' positions can be controlled to move and hold at one of these steps easily with a high degree of accuracy, without any feedback mechanism.

There are a series of electromagnetic coils that can be charged positively or negatively in a specific sequence. Controlling the sequence provides control about the movement, forward or backward in small steps.

Of course, we can do that using Arduino boards.

We are going to examine the unipolar stepper here.

Wiring a unipolar stepper to Arduino

Unipolar steppers usually consist of a center shaft part and four electromagnetic coils. We call them unipolar because power comes in through one pole. We can draw it as follows:

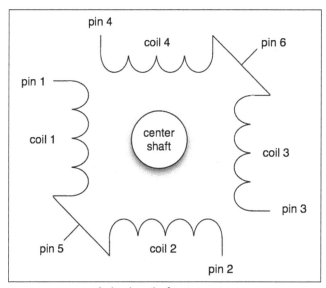

A six-pin unipolar step motor

Let's check how it can be wired to our Arduino.

We need to supply power to the stepper from an external source. One of the best practices here is the use of a wall adapter. Pins 5 and 6 have to be fed a source of current.

Then, we need to control each pin from 1 to 4 with the Arduino. This will be done using the sink current system ULN2004, which is very similar to ULN2003 which we used in the previous chapter with our LED matrix. ULN2004 is suited for voltage from 6 V to 15 V. When ULN2003 is 5 V, the stepper datasheet shows that we have to use this system instead of ULN2003.

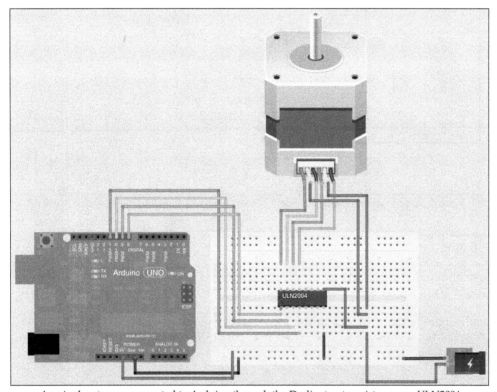

A unipolar stepper connected to Arduino through the Darlington transistor array, ULN2004

Let's check the corresponding circuit diagram:

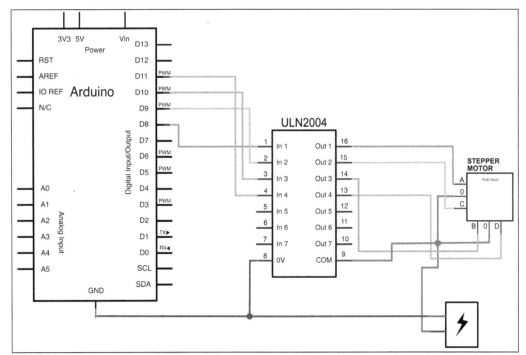

A circuit diagram showing Arduino, the ULN2004 Darlington transistors array, and the stepper

We are using an external power supply here again. All the grounds are wired together too.

Please notice that the **COM** pin (pin number 9) has to be wired to the power supply source (+V).

If you remember correctly from the previous chapter, when we fed an input of the ULN200x Darlington Transistor array, the corresponding output sinks the current to the ground.

In our case here, each pin of Arduino connected to the ULN2004 shift register can commute each pin of the stepper to the ground.

Let's design firmware for stepper control.

Firmware controlling the stepper motor

There is a very nice library that can save us from providing sequences of the HIGH and LOW pins, considering the movements we want to drive.

In order to control precise movements, we normally have to deal with specific sequences. These sequences are usually described in the datasheet.

Let's check the one available at `http://www.sparkfun.com/datasheets/Robotics/StepperMotor.pdf`.

Sparkfun Electronics provides it for a model designed by Robotics.

We can see a table similar to the following one, named **Drive Sequence Model**:

STEP	A	B	C	D
1	HIGH	HIGH	LOW	LOW
2	LOW	HIGH	HIGH	LOW
3	LOW	LOW	HIGH	HIGH
4	HIGH	LOW	LOW	HIGH

If you want to make a clockwise rotation, you should generate a sequence from 1 to 4, then 1, and so on, cyclically. Counterclockwise rotations require generating sequences from 4 to 1 and so on.

Instead of writing a lot of sequences like these, with some function, we can directly use the library named `Stepper`, which is now included in Arduino Core.

Here is the code, followed by the discussion. It is also available in the `Chapter09/StepperMotor/` folder.

```
#include <Stepper.h>
#define STEPS 200

// create an instance of stepper class
Stepper stepper(STEPS, 8, 9, 10, 11);

int counter = 0; // store steps number since last change of direction
int multiplier = 1; // a basic multiplier

void setup()
{
    stepper.setSpeed(30); // set the speed at 30 RPM
}

void loop()
{

    // move randomly from at least 1 step
```

```
    stepper.step(multiplier);

    // counting how many steps already moved
    // then if we reach a whole turn, reset counter and go backward
    if (counter < STEPS)   counter++ ;
    else {
      counter = 0;
      multiplier *= -1;
    }
}
```

We first include the `Stepper` library.

Then we define the number of steps that are equivalent to one whole turn. In our datasheet, we can see that the first step is an angle of 1.8 degrees, with a 5 percent error room. We won't consider that error; we will take 1.8 degrees. This means we need 200 steps (200 * 1.8 = 360°) in order to make a whole turn.

We then instantiate a `Stepper` object by pushing five arguments, which are the step numbers for a whole turn, and the four pins of the Arduino wired to the stepper.

We then declare two helper variables for tracing and, sometimes, changing the rotation direction.

In the `setup()` block, we usually define the speed of the current instance handling the stepper. Here, I have set `30` (which stands for 30 rounds per minute). This can also be changed in the `loop()` block, considering specific conditions or whatever.

At last, in the `loop()` block, we move the stepper to an amount equal to the multiplier value, which is initially `1`. This means that at each run of the `loop()` method, the stepper rotates from step 1 (that is, 1.8 degrees) in the clockwise direction.

I added a logic test, which checks each time if the counter has completed the number of steps required to make a whole turn. If it hasn't, I increment it; otherwise, as soon as it reaches the limit (that is, the motor makes a whole turn since the beginning of the program execution), I reset the counter and invert the multiplier in order to make the stepper continue its walk, but in the other direction.

This is another pattern that you should keep in mind. These are all small patterns that will give you a lot of cheap and efficient ideas to use in each one of your future projects.

With servos and steppers, we can now make things move.

In some of my projects, I used two steppers, with one string bound to each and both these strings bound to a hanging pencil. We can draw on a wall by controlling the amount of string hanging on each side.

Air movement and sounds

Making the air move can generate nice audible sounds, and we are going learn a bit more about this in the following sections.

If you can make things move with Arduino, you will probably be able to make the air move too.

In fact, we have already done this, but we probably didn't move it enough to produce
a sound.

This part is just a short introduction to some definitions and not a complete course about sound synthesis. These are the basic elements that we will use in the next few sections of the book, and as far as possible there will be references of websites or books provided that you can refer to if you are interested in learning more about those specific parts.

What is sound actually?

Sound can be defined as a mechanical wave. This wave is an oscillation of pressure and can be transmitted through solid, liquid, or gas. By extension, we can define sound as the audible result of these oscillations on our ear.

Our ear, combined with further brain processes, is an amazing air-pressure sensor. It is able to evaluate the following:

- Amplitude of a sound (related to the amount of air moving)
- Frequency of a sound (related to the air oscillation amount)

Of course, all these processes are real time, assuming higher or lower frequencies mix at this particular moment.

I'd really suggest that you read the amazing and efficient introduction to *How Digital Audio Works?*, by cycling 74, the maker of the Max 6 framework. You can read it online at `http://www.cycling74.com/docs/max6/dynamic/c74_docs.html#mspdigitalaudio`.

A sound can contain more than one frequency, and it is generally a combination of the frequency content and the global perception of each frequency amplitude that gives the feeling of what we call the timbre of a sound. Psychoacoustics studies the perception of sound.

How to describe sound

We can describe sound in many ways.

Usually, there are two representations of sound:

- Variation of the amplitude over time. This description can be put on a graph and defined as a time-domain representation of sounds.
- Variation of the amplitude depending on the frequency content. This is called the frequency-domain representation of sounds.

There is a mathematical operation that provides an easy way to pass from one to the other, known as the Fourier transform (`http://en.wikipedia.org/wiki/Fast_Fourier_transform`). Many implementations of this operation are available on computers, in the form of the **Fast Fourier Transform** (**FFT**), which is an efficient method that provides fast approximate calculations.

Let's consider a sinusoidal variation of air pressure. This is one of the most simple sound waves.

Here are the two representations in the two domains:

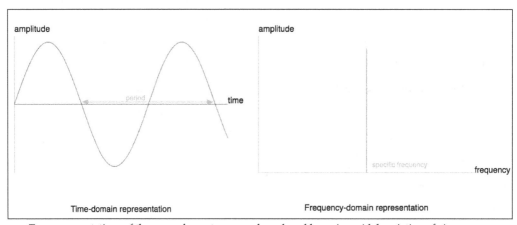

Two representations of the same elementary sound produced by a sinusoidal variation of air pressure

Let's describe the two graphs of the preceding image.

In the time-domain representation, we can see a cyclical variation with a period. The period is the time equivalent of the spatial wavelength.

The period is the time needed to complete a complete vibrational cycle. Basically, if you can describe the variation over a period, you are able to totally draw the representation of the sound in time. Here, it is a bit obvious because we are watching a pure sine-based sound.

If you draw and observe a sound produced by a source, the amplitude variation over time will correspond directly to a variation of air pressure.

Considering the orientation of the axis, we first have what we call a high-pressure front. This is the part of the curve above zero (represented by the time axis). This means that the pressure is high and our tympanum is pushed a bit more inside our ear.

Then, after a semi-period, the curve crosses zero and goes below, meaning that the air pressure is lower than the normal atmospheric pressure. Our tympanum also feels this variation. It is pulled a little bit.

In the frequency-domain representation, there is only a vertical line. This pulse-like graph in the previous figure represents the unique frequency contained in this sine-based sound. It is directly related to its period by a mathematical equation, as follows:

$$T = \frac{1}{f}$$

Here, T is the period in seconds and f is the frequency in Hertz.

The higher the frequency, the more the sound is felt as high-pitched. The lesser it is, the more the sound is felt as low-pitched.

Of course, a high frequency means a short period and faster oscillations over time.

These are the basic steps in understanding how sound can be represented and felt.

Microphones and speakers

Microphones are devices that are sensitive to the subtle variation of air pressure. Yes, they are sensors. They can translate air-pressure variations into voltage variations.

Speakers are devices that implement a part that can move, pushing and pulling masses of air, making it vibrate and produce sounds. The movement is induced by voltage variations.

In both these cases, we have:

* A membrane
* An electrical transducer system

In the microphone case, we change the air pressure and that produces an electrical signal.

In the speaker case, we change the electrical signal and that produces an air pressure variation.

In each case, we have analog signals.

Digital and analog domains

Sounds sources can be very different. If you knock on a table, you'll hear a sound. This is a basic analog- and physical-based sound. Here, you physically make the table vibrate a bit, pushing and pulling air around it; and because you are near it, your tympanum feels these subtle variations.

As soon as we talk about digital equipment, we have some limitations considering storage and memory. Even if these are large and sufficient now, they aren't infinite.

And how can we describe something analog in that case? We already spoke about this situation when we described analog and digital input and output pins of Arduino.

How to digitalize sound

Imagine a system that could sample the voltage variation of your microphones periodically. A sampling concept usually used is sample and hold.

The system is able to read the analog value at regular intervals of time. It takes a value, holds it as a constant until the next value, and so on.

We are talking about the sampling rate to define the sampling frequency. If the sampling rate is low, we will have a lower approximation of the analog signal than if what we would have had if the sampling rate was high.

A mathematical theorem provides us a limit that we have to keep in mind — the Nyquist frequency.

In order to keep our sampling system process a safe artifact induced by the system itself, we have to sample at a minimum of two times the higher frequency in our original analog signal.

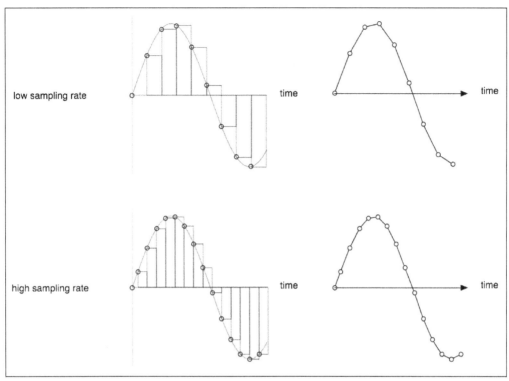

Example illustrating the sampling rate while sampling a sine wave

A higher sampling rate not only means more precision and fidelity to the original analog wave, but also more points to store in the digital system. The result would be a heavier file, in terms of disks and filesystems.

Another element to keep in mind while sampling is the bit depth.

I voluntarily omitted it in the previous figure in order to not overload the drawings.

Indeed, we sampled a value over time, but how can you represent the value itself, the amplitude I mean? We use a bit-based coding system, as usual, with the digital equipment.

The **bit depth** is the resolution of the amplitude values from -1 (the minimum possible) to 1 (the maximum possible).

The higher the bit depth, the more the subtle variations we can encode and record into our digital systems. Conversely, if we have a very low bit-depth sampler and we make a progressively decreasing amplitude variation, the sound will decrease considerably in a manner similar to the Doppler effect. For instance, we wouldn't be able to distinguish values from 0.5 to 0.6; everything would only be 0.5 or 0.7 but never 0.6. The sound would lose subtlety.

Usual sampling rates and bit depth depends on the purpose of the final rendering.

Here are two commonly used quality standards:

- CD quality is 44.1 kHz and 16-bit
- DAT quality is 48 kHz and 16-bit

Some recording and mastering studios use audio interfaces and internal processing at 96 kHz and 24 bits. Some people who love old-school sound engines still use lo-fi systems to produce their own sound and music at 16 kHz and 8 bits.

The process from analog to digital conversion is handled by the **analog to digital converter (ADC)**. Its quality is the key to achieving good conversion. This process is similar to the one involved in Arduino when we use an analog input. Its ADC is 10 bits and it can read a value once every 111 microseconds, which is a sampling rate frequency of 9 kHz.

Buffers are used to smoothly process times and make things smoother in time.

How to play digital bits as sounds

We can also convert digital encoded sounds into analog sounds. This process is achieved by the **digital to analog converter (DAC)**.

If the processor sends bits of data from the encoded sound to the DAC as a continuous flow of discrete values, the DAC takes all these values and converts them as an analog electrical signal. It interpolates values between each digital value, which often involves some processes (for example, low-pass filtering), in order to remove some artifacts such as harmonics above the Nyquist frequency.

In the world of digital audio, DAC power and quality is one of the most important aspects of our audio workstation. They have to provide high resolutions, a high sampling rate, a small total harmonic distortion and noise, and a great dynamic range.

How Arduino helps produce sounds

Let's come back to Arduino.

Arduino can read and write digital signals. It can also read analog signals and simulate analog output signals through PWM.

Wouldn't it be able to produce and even listen to sounds? Of course it would.

We can even use some dedicated components to make things better. For instance, we can use an ADC with a higher sampling rate in order to store sounds and a high-quality DAC too, if required. Today, we often use electronic hardware equipment to control software. We can, for instance, build a device based on Arduino, full of knobs and buttons and interface it with a software on the computer. This has to be mentioned here.

We can also use Arduino as a sound trigger. Indeed, it is quite easy to turn it into a small sequencer, popping out specific MIDI or OSC messages to an external synthesizer, for instance. Let's move further and go deeper into audio concepts specifically with the Arduino board.

Playing basic sound bits

Playing a sound requires a sound source and a speaker. Of course, it also requires a listener who is able to hear sounds.

Natively, Arduino is able to produce 8 kHz and 8-bit audio playback sounds on small PC speakers.

We are going to use the `tone()` function available natively in the Arduino Core. As written at `http://arduino.cc/en/Reference/Tone`, we have to take care of the pins used when using this function, because it will interfere with PWM output on pins 3 and 11 (except for the Arduino MEGA).

This technique is also named **bit-banging**. It is based on I/O pin toggling at a specific frequency.

Wiring the cheapest sound circuit

We are going to design the cheapest sound generator ever with a small 8-ohm speaker, a resistor, and an Arduino board.

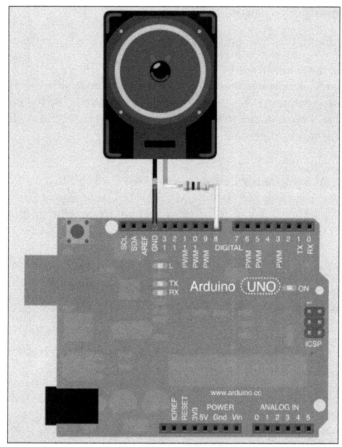

A small sound generator

The connections made here ensure an audible sound. Let's program the chip now.

The corresponding circuit diagram is as follows:

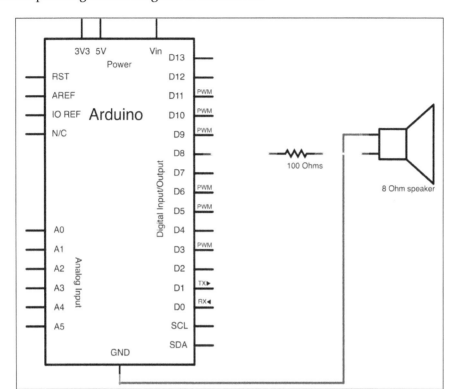

The diagram of the sound generator

Playing random tones

As a digital artist and specifically as an electronic musician, I like to be free of the notes. I often use frequencies instead of notes; if you are interested, you can read about the microtonal concept at http://en.wikipedia.org/wiki/Microtonal_music.

In this example, we don't use notes but frequencies to define and trigger our electronic music.

The code is also available in the Chapter09/ ToneGenerator/ folder.

```
void setup() {
  // initiate the pseudo-random number generator
  randomSeed(analogRead(0));
}

void loop() {
```

```
    // generate random pitch & duration
    int pitch = random(30,5000);
    int duration = 1000 / (random(1000) + 1);

    // play a tone to the digital pin PWM number 8
    tone(8, pitch, duration);

    // make a pause
    delay(duration * 1.30);

    // stop the tone playing
    noTone(8);
}
```

We initialize the pseudorandom number generator at first by reading the analog input 0.

In the loop, we generate two numbers:

- The pitch is a number from 30 to 4,999; this is the frequency of the sound
- The duration is a number from 1 ms to 1 s; this is the duration of the sound

These two arguments are required by the tone() function.

Then, we call tone(). The first argument is the pin where you feed the speaker.

The tone() function generates a square wave of the specified frequency on a pin as explained in its reference page at http://arduino.cc/en/Reference/Tone.

If we don't provide a duration, the sound continues until the noTone() function is called. The latter takes an argument that was used by the pin as well.

Now, listen to and enjoy this microtonal pseudorandom melody coming from your 8-bit chip.

Improving the sound engine with Mozzi

The bit-banging technique is very cheap and it's nice to learn how it works. However, I can quote some annoying things here:

- **No pure sound**: Square waves are a sum of all odd harmonics at the fundamental frequency
- **No amplitude control available**: Each note sounds at the same volume

We are going to use a very nice library called Mozzi, by Tim Barrass. The official website is directly hosted on GitHub at `http://sensorium.github.com/Mozzi/`. It includes the `TimerOne` library, a very fast timer handler.

Mozzi provides a very nice 16,384 kHz, 8-bit audio output. There is also a nice basic audio toolkit containing oscillators, samples, lines and envelopes, and filtering too.

Everything is available without external hardware and by only using two pins of the Arduino.

We are going to design a small sound engine based on it.

Setting up a circuit and Mozzi library

Setting up the circuit is easy; it is the same as the latest one except that pin 9 has to be used.

Mozzi's documentation says:

> *To hear Mozzi, connect a 3.5 mm audio jack with the centre wire to the PWM output on Digital Pin 9* on Arduino, and the black ground to the Ground on the Arduino. Use this as a line out which you can plug into your computer and listen to with a sound program like Audacity.*

> *It is really easy to set up the hardware. You can find many 3.5 mm audio jack connector like that all over the Internet. In the following circuit diagram, I put a speaker instead of a jack connector but it works exactly the same with a jack connector, that latter having 2 pins, one ground and one signal related. Ground has to be connected to the Arduino's ground and the other pin to the digital pin 9 of the Arduino.*

> *Then we have to install the library itself.*

> *Download it from their website: http://sensorium.github.com/Mozzi*

> *Unzip it and rename the folder as Mozzi.*

> *Then put it as usual in the place you put your libraries; in my case it is:*

> */Users/julien/Documents/Arduino/libraries/*

> *Restart or just start your Arduino IDE and you'll be able to see the library in the IDE.*

> *It is provided with a bunch of examples.*

> *We are going to use the one about the sine wave.*

This is what the Mozzi library looks like:

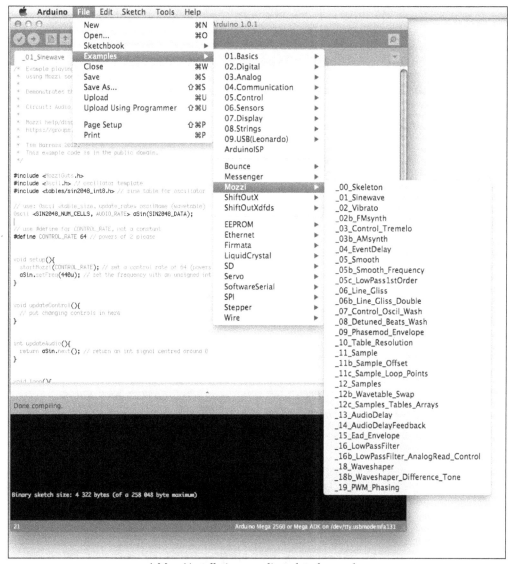

A Mozzi installation revealing a lot of examples

An example sine wave

As with any library, we have to learn how to use the sine wave.

There are a lot of examples, and these are useful to learn how to design our own firmware step-by-step. Obviously, I won't describe all these examples, but only those in which I'll grab elements to make your own sound generator.

Let's check the sine wave example. It is also available in the `Chapter09/MozziSoundGenerator/` folder.

```
#include <MozziGuts.h>
#include <Oscil.h> // oscillator template
#include <tables/sin2048_int8.h> // sine table for oscillator

// use: Oscil <table_size, update_rate> oscilName (wavetable)
Oscil <SIN2048_NUM_CELLS, AUDIO_RATE> aSin(SIN2048_DATA);

// use #define for CONTROL_RATE, not a constant
#define CONTROL_RATE 64 // powers of 2 please


void setup(){
  startMozzi(CONTROL_RATE); // set a control rate of 64 (powers of 2
please)
  aSin.setFreq(440u); // set the frequency with an unsigned int or a
float
}


void updateControl(){
  // put changing controls in here
}


int updateAudio(){
  return aSin.next(); // return an int signal centered around 0
}


void loop(){
  audioHook(); // required here
}
```

At first, some inclusions are done.

`MozziGuts.h` is the basic header to include in any case.

`Oscil.h` is the header to use if you need an oscillator.

We then include a wave table (sine wave).

Oscillators

In the sound synthesis world, an **oscillator** is a basic unit that is capable of producing oscillations. It is often used not only for direct sound generation with frequencies varying from 20 Hz to 20 kHz (audible spectrum), but also as a modulator (usually with frequencies lower than 50 Hz). It has been used as the latter in this case. An oscillator is usually called a **Low Frequency Oscillator (LFO)**.

Wavetables

A **wavetable** is a very nice and efficient way to store whole pieces of sounds, generally cyclical or looped sounds.

We basically used this as a lookup table. Do you remember using it?

Instead of calculating our sine value over time in real time, we basically precalculate each value of a whole period, and then add the results into a table; each time we need it, we just have to scan the table from the beginning to the end to retrieve each value.

Of course, this IS definitely an approximation. But it saves a lot of CPU work.

A wavetable is defined by its size, the sample rate related, and of course the whole values.

Let's check what we can find in the `sin2048_int8.h` file:

```c
#ifndef SIN2048_H_
#define SIN2048_H_

#include "Arduino.h"
#include <avr/pgmspace.h>

#define SIN2048_NUM_CELLS 2048
#define SIN2048_SAMPLERATE 2048

const char __attribute__((progmem)) SIN2048_DATA [] =
        {
                0, 0, 0, 1, 1, 1, 2, 2, 3, 3, 3, 4, 4, 5,
                5, 5, 6, 6, 7, 7, 7, 8, 8, 9, 9, 9, 10, 10, 10, 11, 11, 12, 12, 12, 13, 13, 14,
                14, 14, 15, 15, 16, 16, 16, 17, 17, 18, 18, 18, 19, 19, 19, 20, 20, 21, 21, 21,
                22, 22, 23, 23, 23, 24, 24, 24, 25, 25, 26, 26, 26, 27, 27, 28, 28, 28, 29, 29,
                29, 30, 30, 31, 31, 31, 32, 32, 33, 33, 34, 34, 35, 35, 36, 36, 36, 37,
                37, 37, 38, 38, 39, 39, 39, 40, 40, 40, 41, 41, 42, 42, 42, 43, 43, 43, 44, 44,
                44, 45, 45, 46, 46, 46, 47, 47, 47, 48, 48, 48, 49, 49, 50, 50, 50, 51, 51, 51,
                52, 52, 52, 53, 53, 54, 54, 54, 55, 55, 55, 56, 56, 56, 57, 57, 57, 58, 58, 58,
                59, 59, 59, 60, 60, 61, 61, 61, 62, 62, 62, 63, 63, 63, 64, 64, 64, 65, 65, 65,
                66, 66, 66, 67, 67, 67, 68, 68, 68, 69, 69, 69, 70, 70, 70, 71, 71, 71, 72, 72,
                72, 73, 73, 73, 74, 74, 74, 74, 75, 75, 75, 76, 76, 76, 77, 77, 77, 78, 78, 78,
                79, 79, 79, 79, 80, 80, 80, 81, 81, 81, 82, 82, 82, 83, 83, 83, 83, 84, 84, 84,
                85, 85, 85, 85, 86, 86, 86, 87, 87, 87, 88, 88, 88, 89, 89, 89, 90, 90,
                90, 91, 91, 91, 91, 92, 92, 92, 92, 93, 93, 93, 94, 94, 94, 94, 95, 95, 95, 95,
                96, 96, 96, 96, 97, 97, 97, 97, 98, 98, 98, 98, 99, 99, 99, 99, 100, 100, 100,
                100, 101, 101, 101, 101, 102, 102, 102, 102, 103, 103, 103, 103, 103, 104, 104,
                104, 104, 105, 105, 105, 105, 105, 106, 106, 106, 106, 107, 107, 107, 107, 107,
                108, 108, 108, 108, 108, 109, 109, 109, 109, 109, 110, 110, 110, 110, 110, 111,
                111, 111, 111, 111, 112, 112, 112, 112, 112, 113, 113, 113, 113, 113, 113, 114,
```

We can indeed find the number of cells: 2048 (that is, there are 2048 values in the table). Then, the sample rate is defined as 2048.

Let's go back to the example.

We then define the Oscil object that creates an oscillator.

After the second `define` keyword related to the variable update frequency, we have the usual structure of `setup()` and `loop()`.

We also have `updateControl()` and `updateAudio()` and those aren't defined in the code. Indeed, they are related to Mozzi and are defined in the library files themselves.

The `setup()` block starts the Mozzi library at the specific control rate defined before. Then, we set up the oscillator defined before at a frequency of 440 Hz. 440 Hz is the frequency of the universal A note. In this context, it can be thought of as the audio equivalent of the Hello World example.

Nothing more about `updateControl()` here.

We return `aSin.next()` in `updateAudio()`. It reads and returns the next sample, which is understood as the next element, which is the next bit of sound.

In `loop()`, we call the `audioHook()` function.

The global pattern is usual. Even if you use another library related to sound, inside or outside the Arduino world, you'll have to deal with this kind of pattern in four steps (generally, but it may differ):

- Definitions in the header with some inclusions
- Start of the audio engine
- Permanent loop of a hook
- Updating functions for rendering things before a commit, then in the hook

If you upload this, you'll hear a nice A440 note, which may make you hum a little.

Frequency modulation of a sine wave

Let's now merge some concepts — sine wave generation, modulation, and input reading.

We are going to use two oscillators, one modulating the frequency of the other.

With a potentiometer, we can control the frequency of the modulating oscillator.

Let's first improve the circuit by adding a potentiometer.

Adding a pot

In the following circuit diagram, we have added a potentiometer in the sound generator circuit:

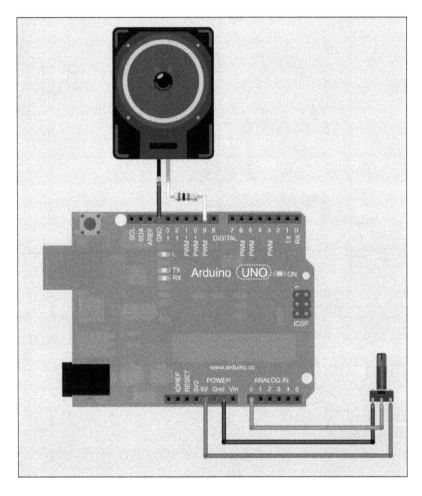

The circuit diagram is as follows:

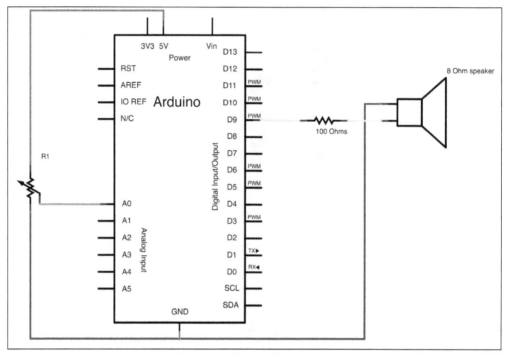

Improving the sound generator

Upgrading the firmware for input handling

This code is also available in the Chapter09/MozziFMOnePot/ folder.

```
#include <MozziGuts.h>
#include <Oscil.h>
#include <tables/cos8192_int8.h> // table for Oscils to play
#include <utils.h> // for mtof

#define CONTROL_RATE 64 // powers of 2 please

Oscil<COS8192_NUM_CELLS, AUDIO_RATE> aCos(COS8192_DATA);
Oscil<COS8192_NUM_CELLS, AUDIO_RATE> aVibrato(COS8192_DATA);

const long intensityMax = 500;

int potPin = A0;
int potValue = 0;
```

```
void setup(){
  startMozzi(CONTROL_RATE);
  aCos.setFreq(mtof(random(21,80)));
  aVibrato.setFreq((float) map(potValue, 0, 1024, 0, intensityMax));
}

void loop(){
  audioHook();
}

void updateControl(){
  potValue = analogRead(potPin);
}

int updateAudio(){
  long vibrato = map(potValue, 0, 1024, 0, intensityMax) * aVibrato.
next();
  return (int)aCos.phMod(vibrato);
}
```

In this example, we use two oscillators, both based on a cosine wavetable:

- aCos stands for the sound itself
- aVibrato is the modulator

Since we have a potentiometer here, we need to scale things a bit.

intensityMax is the maximum intensity of the modulation effect. I chose 500 after testing it myself.

We often use the following technique to scale things: use a constant (or even a "real" variable) and then multiply it by the value you can vary. This can be done in one pass by using the map() function. We already used it in *Chapter 6, Sensing the World–Feeling with Analog Inputs*, for the same purpose—scaling an analog input value.

In that case, at the maximum value, your potentiometer (more generally your input) changes the parameter you want to alter to its maximum value.

Let's continue the review of the code.

We define the potentiometer pin n and the variable potPin. We also define potValue to 0.

In the setup() block, we start Mozzi. We define the frequency of the oscillator as aCos. The frequency itself is the result of the mtof() function. mtof stands for **MIDI to Frequency**.

As we are going to describe it a bit later, MIDI protocol codes many bytes of values, including the pitch of notes it uses to transport from sequencers to instruments. Each MIDI note fits with real note values in the real world, and each note fits with a particular frequency. There are tables that show the frequency of each MIDI note, and Mozzi includes that for us.

We can pass a MIDI note pitch as argument to the `mtof()` function, and it will return the right frequency. Here, we use the `random(21,80)` function to generate a MIDI note pitch from 21 to 79, which means from A0 to A5.

Of course, this use case is a pretext to begin introducing MIDI. We could have directly used a `random()` function to generate a frequency.

We then read the current value of the analog input A0 and use it to calculate a scaled value of the frequency of the modulating oscillator, `aVibrato`. This is only to provide more randomness and weirdness. Indeed, if your pot isn't at the same place each time you restart Arduino, you'll have a different modulation frequency.

The `loop()` block then executes the `audioHook()` method constantly to produce audio.

And the smart thing here is the `updateControl()` method. We add the `analogRead()` function that reads the value of the analog input. Doing this in `updateControl()` is better, considering the purpose of this function. Indeed, the Mozzi framework separates the audio rendering time-critical tasks from the control (especially human control) pieces of code.

You'll come across this situation very often in many frameworks, and it can confuse you the first time. It is all about the task and its scheduling. Without reverse-engineering the Mozzi concepts here, I would like to say only that time-critical events have to be handled more carefully than human actions.

Indeed, even if it seems as if we can be very fast at turning a knob, it is really slow compared to the sample rate of Mozzi, for instance (16,384 kHz). This means we cannot stop the whole process only to test and check, if we change the value of this potentiometer constantly. Things are separated; keep this in mind and use the framework carefully.

Here, we read the value in `updateControl()` and store it in the `potValue` variable.

Then, in `updateAudio()`, we calculate the vibrato value as the value of `potValue` scaled from `0` to the value of `intensityMax`, multiplied by the next value of the oscillator in its wavetable.

This value is then used in a new method named `phMod`. This method applies a phase modulation to the oscillator for which it is called. This modulation is a nice way to produce a frequency modulation effect.

Now, upload the firmware, add the earphone, and turn the potentiometer. You should be able to hear the effect and control it with the potentiometer.

Controlling the sound using envelopes and MIDI

We are now okay to design small bits of a sound engine using Mozzi. There are other libraries around, and what we learned will be used with those two. Indeed, these are patterns.

Let's check how we can control our Arduino-based sound engine using a standard protocol from a computer or other device. Indeed, it would be interesting to be able to trigger notes to change sound parameters using a computer, for instance.

Both are protocols used in the music and new media related projects and works.

An overview of MIDI

MIDI is short for **Musical Instrument Digital Interface**. It is a specification standard that enables digital music instruments, computers, and all required devices to connect and communicate with one another. It was introduced in 1983, and at the time of writing has just celebrated its 30th anniversary. The reference website is `http://www.midi.org`.

MIDI can transport the following data over a basic serial link:

- Notes (on/off, after touch)
- Parameter changes (control change, program change)
- Real-time messages (clock, transport state such as start/stop/continue)
- System exclusives, allowing manufacturers to create their message

A new protocol appeared and is used very widely today: OSC. It isn't a proper protocol, by the way.

OSC stands for **Open Sound Control** and is a content format developed by two people at the **Center for New Music and Audio Technologies (CNMAT)** at University of Berkeley, California. It was originally intended for sharing gestures, parameters, and sequences of notes during musical performances. It is very widely used as a replacement for MIDI today, providing a higher resolution and faster transfer. Its main feature is the native network transport possibility. OSC can be transported over UDP or TCP in an IP environment, making it easy to be used over Wi-Fi networks and even over the Internet.

MIDI and OSC libraries for Arduino

I'd suggest two libraries here. I tested them myself and they are stable and efficient. You can check the one about MIDI at `http://sourceforge.net/projects/arduinomidilib`. You can check this one about OSC at `https://github.com/recotana/ArdOSC`. You shouldn't have too many difficulties installing them now. Let's install at least MIDI, and restart the IDE.

Generating envelopes

In the audio field, an **envelope** is a shape used to modify something. For instance, imagine an amplitude envelope shaping a waveform.

You have a waveform first. I generated this sine with Operator synthesizer in Ableton Live (`https://www.ableton.com`), the famous digital audio workstation. Here is a screenshot:

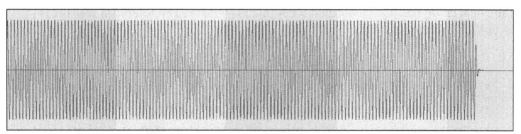

A basic sine wave generated by an operator in Ableton Live's Operator FM synth

The sine doesn't show very well due to aliasing; here is another screenshot, which is the same wave but more zoomed in:

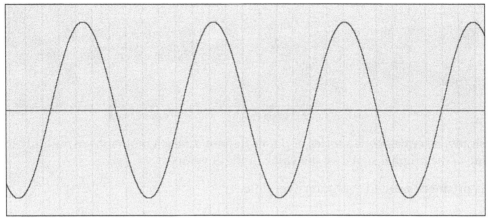

A sine wave

This sine wave has a global constant amplitude. Of course, the air pressure push and pull constantly, but the global maximums and minimums are constant over time.

Musicians always want to make their sounds evolve over time, subtly or harshly.

Let's apply an envelope to this same wave that will make it increase the global volume progressively, then decrease it a bit, and then decrease quickly to zero:

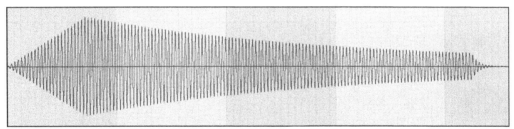

A sine wave altered by an envelope with a long attack

Here is the result with another envelope:

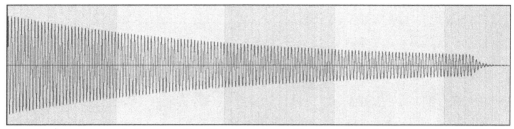

A sine wave altered by an envelope with a very short attack

Basically, an envelope is a series of points in time. At each moment, we multiply the value of the original signal by the value of the envelope.

This produces a sound evolution over time.

We can use envelopes in many cases because they can modulate amplitude, as we just learned. We can also use them to alter the pitch (that is, the frequency) of a sound.

Usually, envelopes are triggered (that is, applied to the sound) at the same time the sound is triggered, but of course we can use the offset retrigger feature to retrigger the envelope during the same triggered sound and do much more.

Here is a last example showing a pitch envelope. The envelope makes the frequency of the sound decrease. As you can see, the waves are tighter on the left than on the right. The sound changes from high-pitched to low-pitched.

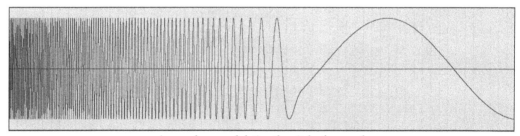

An envelope modulating the pitch of a sound

Implementing envelopes and MIDI

We are going to design a very cheap sound synthesizer that will be able to trigger notes when it receives a MIDI note message and alter the sound when it receives a particular MIDI Control Change message.

The MIDI part will be handled by the library and the envelope will be explicated and coded.

You can check the following code. This code is also available in the Chapter09/MozziMIDI/ folder.

```
#include <MIDI.h>
#include <MozziGuts.h>
#include <Oscil.h> // oscillator template
#include <Line.h> // for envelope
#include <utils.h> // for mtof
#include <tables/sin2048_int8.h> // sine table for oscillator


// use #define for CONTROL_RATE, not a constant
#define CONTROL_RATE 128 // powers of 2 please

// declare an oscillator using a sine tone wavetable
// use: Oscil <table_size, update_rate> oscilName (wavetable)
Oscil <SIN2048_NUM_CELLS, AUDIO_RATE> aSin(SIN2048_DATA);

// for envelope
Line <unsigned int> aGain;
unsigned int release_control_steps = CONTROL_RATE; // 1 second of
control
unsigned int release_audio_steps = 16384; // 1 second of audio
int fade_counter;

float vol= 1. ; // store the master output volume

unsigned int freq; // to convey control info from MIDI handler to
updateControl()


void HandleControlChange(byte channel, byte CCnumber, byte value) {
  switch(CCnumber){
    case 100:
      vol = map(value,0, 127, 0., 1.);
    break;
  }
}

void HandleNoteOn(byte channel, byte pitch, byte velocity) {
```

```
    // scale velocity for high resolution linear fade on Note-off later
    freq = mtof(pitch);
    aGain.set(velocity<<8); // might need a fade-in to avoid clicks


}


void HandleNoteOff(byte channel, byte pitch, byte velocity) {

    // scale velocity for high resolution linear fade on Note-off later
    aGain.set(0,release_audio_steps);
    fade_counter = release_control_steps;
}

void setup() {

    // Initiate MIDI communications, listen to all channels
    MIDI.begin(MIDI_CHANNEL_OMNI);

    // Connect the HandleControlChange function to the library, so it is
    called upon reception of a NoteOn.
    MIDI.setHandleControlChange(HandleControlChange); // Put only the
    name of the function

        // Connect the HandleNoteOn function to the library, so it is
    called upon reception of a NoteOn.
    MIDI.setHandleNoteOn(HandleNoteOn);  // Put only the name of the
    function


        // Connect the HandleNoteOn function to the library, so it is
    called upon reception of a NoteOn.
    MIDI.setHandleNoteOff(HandleNoteOff);  // Put only the name of the
    function

    aSin.setFreq(440u); // default frequency
    startMozzi(CONTROL_RATE);
}


void updateControl(){
    // Ideally, call MIDI.read the fastest you can for real-time
performance.
    // In practice, there is a balance required between real-time
```

```
    // audio generation and a responsive midi control rate.
    MIDI.read();

    if (fade_counter-- <=0) aGain.set(0,0,2); // a line along 0
}


int updateAudio(){
    // aGain is scaled down to usable range
    return (int) ((aGain.next()>>8) * aSin.next() * vol )>>8; // >>8
shifts the multiplied result back to usable output range
}


void loop() {
    audioHook(); // required here
}
```

At first, we include the MIDI library. Then we include the Mozzi library.

Of course, the right bits of Mozzi to include are a bit different for each project. Studying examples helps to understand what goes where. Here, we not only need Oscil for the basic features of the oscillator, but also need Line. Line is related to interpolation functions in Mozzi. Generating an envelope deals with this. Basically, we choose two values and a time duration, and it starts from the first one and reaches the second one in the time duration you choose.

We also include the wavetable related to a sine.

We define a control rate higher than before, at 128. That means the updateControl() function is called 128 times per second.

Then we define the oscillator as aSin.

After these bits, we define an envelope by declaring an instance of the Line object.

We define two variables that store the release part of the envelope duration, one for the control part in one second (that is, the number of steps will be the value of CONTROL_RATE) and one for the audio part in one second too (that is, 16,384 steps). Lastly, a variable named fade_counter is defined.

`HandleControlChange()` is a function that is called when a MIDI Control Change message is sent to Arduino. The message comes with these bytes:

- MIDI channel
- CC number
- Value

These arguments are passed to the `HandleControlChange()` function, and you can access them directly in your code.

This is a very common way to use event handlers. Almost all event listener frameworks are made like this. You have some function and you can use them and put whatever you want inside them. The framework itself handles the functions that have to be called, saving as much CPU time as possible.

Here, we add a `switch` statement with only one case over the `CCNumber` variable.

This means if you send a MIDI Control Change 100 message, this case being matched, the value of `CC` will be processed and the `vol` variable will be altered and modified. This Control Change will control the master output volume of the synth.

In the same way, `HandleNoteOn()` and `HandleNoteOff()` handle MIDI note messages.

Basically, a MIDI Note On message is sent when you push a key on your MIDI keyboard. As soon as you release that key, a MIDI Note Off message pops out.

Here, we have two functions handling these messages.

`HandleNoteOn()` parses the message, takes the velocity part, bit shifts it on the left to 8 bits, and passes it to `aGain` through the `set()` method. When a MIDI Note On message is received, the envelope `aGain` is triggered to its maximum value. When a MIDI Note Off message is received, the envelope is triggered to reach 0 in one second via the number of audio steps discussed before. The `fade` counter is also reset to its maximum value at the moment the key is released.

In this way, we have a system responding to the MIDI Note On and MIDI Note Off messages. When we push a key, a sound is produced until we release the key. When we release it, the sound decreases linearly to 0, taking one second.

The `setup()` method includes the setup of the MIDI library:

- `MIDI.begin()` instantiates the communication
- `MIDI.setHandleControlChange()` lets you define the name of the function called when a control change message is coming
- `MIDI.setHandleNoteOn()` lets you define the name of the function called when a Note On message is coming
- `MIDI.setHandleNoteOff()` lets you define the name of the function called when a Note Off message is coming

It also includes the setup of Mozzi.

The `loop()` function is quite familiar now.

The `updateControl()` function does not contain the time-critical part of the sound generator. It doesn't mean this function is called rarely; it is called less than `updateAudio()` — 128 times per second for control and 16,384 per second for audio, as we have seen before.

This is the perfect place to read our MIDI flow, with the `MIDI.read()` function.

This is where we can trigger our decreasing envelope to 0 as soon as the `fade` counter reaches 0 and not before, making the sound in one second, as we checked before.

Lastly, the `updateAudio()` function returns the value of the oscillator multiplied by the envelope value too. This is the purpose of the envelope. Then, `vol` multiplies the first result in order to add a key to control the master output volume.

The `<<8` and `>>8` expressions here are for setting a high-resolution linear fade on Note Off, and this is a nice trick provided by Tim Barrass himself.

Wiring a MIDI connector to Arduino

This schematic is based on the MIDI electrical specification diagram at
http://www.midi.org/techspecs/electrispec.php.

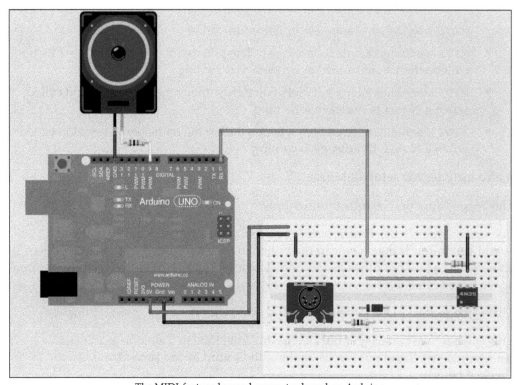

The MIDI-featured sound generator based on Arduino

The corresponding circuit diagram is as follows:

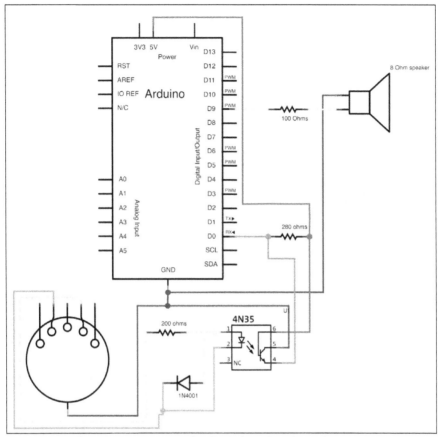

The MIDI connector wired to the Arduino-based sound generator

As you can see, the digital pin 0 (serial input) is involved. This means we won't be able to use the serial communication over USB. In fact, we want to use our MIDI interface.

Let's upload the code and start this small sequencer in Max 6.

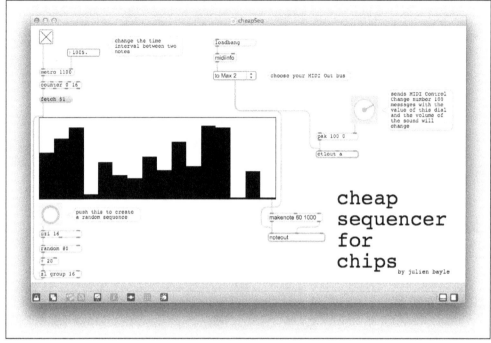

The *cheap sequencer for chips* fires MIDI notes and MIDI control changes

The sequencer is quite self-explanatory. Toggle on the toggle button at the top-left and it starts the sequencer, reading each step in the multislider object. The higher a slider is, the higher the pitch of this note into that step will be.

You can click on the button under the multislider on the left, and it will generate a random sequence of 16 elements.

Choose the correct MIDI output bus from the list menu on the top-right.

Connect your Arduino circuit and your MIDI interface with a MIDI cable, and listen to the music. Change the multislider content and the sequence played. If you turn the dial, the volume will change.

Everything here is transmitted by MIDI. The computer is a sequencer and a remote controller and the Arduino is the synthesizer.

Playing audio files with the PCM library

Another way to play sounds is by reading already digitalized sounds.

Audio samples define digital content, often stored as files on filesystems that can be read and converted into audible sound.

Samples can be very heavy from the memory size point of view.

We are going to use the PCM library set up by David A. Mellis from MIT. Like other collaborators, he is happy to be a part of this book.

The reference page is `http://hlt.media.mit.edu/?p=1963`.

Download the library and install it.

Imagine that we have enough space in the Arduino memory spaces. How can we do the installation if we want to convert a sample on our disks as a C-compatible structure?

The PCM library

Check this code. It is also available in the `Chapter09/PCMreader/` folder.

Our PCM reader

There is an array of `unsigned char` datatypes declared as `const`, and especially with the `PROGMEM` keyword named `sample`.

`PROGMEM` forces this constant to be put in the program space instead of RAM, because the latter is much smaller. Basically, this is the sample. The `startPlayback()` function is able to play a sample from an array. The `sizeof()` method calculates the size of the memory of the array.

WAV2C – converting your own sample

Since we have already played with wavetable, and this is what we will be doing hereafter, we can store our sample waveforms in the Arduino code directly.

Even if dynamic reading of the audio file from an SD card would seem smarter, PCM provides an even easier way to proceed — directly reading an analog conversion of an array while storing a waveform into a sound.

We first have to transform a sample as C data.

David Ellis made an open source, small processing-based program that provides a way to do this; it can be found at `https://github.com/damellis/EncodeAudio`.

You can download it from the reference project page directly compiled for your OS.

Launch it, choose a WAV file (PCM-based encoded sample), and then it will copy something huge in your clipboard.

Then, you only have to copy-paste this content into the array defined before.

Be careful to correctly paste it between the curly brackets.

Here is the content copied from the clipboard after converting a `wav` sample that I made myself:

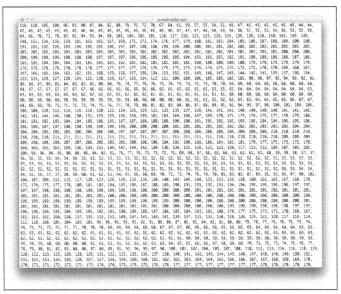

A huge amount of data to paste in a C array

In the same folder, I have put a `.wav` file I designed. It is a short rhythm recorded in 16 bits.

Wiring the circuit

The circuit is similar to the one in the *Playing basic sound bits* section, except that we have to use the digital pin 11 here. And we cannot use PWM on pins 3, 9, and 10 because the timers involved in the library consume them.

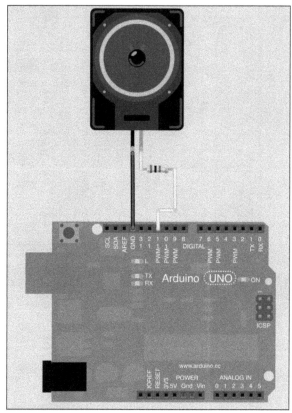

Wiring our PCM reader

The circuit diagram is easy too.

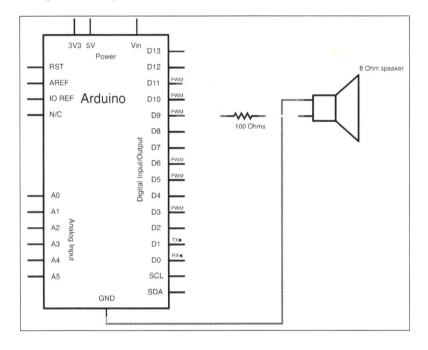

Don't forget to use pin 11 with the PCM library

Now, let's play the music.

Other reader libraries

There are also other libraries providing ways to read and decode the MP3 format or other formats.

You can find a lot on the Internet; but be careful as some of them require some shields, like the one on the Sparkfun website at https://www.sparkfun.com/products/10628.

This provides a shield with an SD Card reader, a 3.5 mm stereo headphone jack, a VS1053 shift register, and very versatile decoder chips for MP3, WMA, AAC, and other formats.

It is a very dedicated solution and we only have to interface the shield with Arduino.

Arduino only sends and receives bits from the shield, which takes care of the decoding of the encoded files, the conversion to analog signals, and so on.

I'd really suggest testing it. There are many examples on the Sparkfun website.

Summary

We learned how to make things move right here with Arduino. In particular, we learned about:

- Moving solid things with motors
- Moving air with sound generators

Of course, unfortunately, I cannot describe more on how to make things move.

If you need help with sound, please contact me at book@cprogrammingforarduino.com. I will be a happy to help you with sound inputs too, for instance.

This is the end of the second part of the book. We discovered a lot of concepts together. And now we are going to dig into some more advanced topics.

We are able to understand firmware design and inputs and outputs, so let's move further.

We are going to dig deeper into precise examples with I2C/SPI communication to use GPS modules, 7-segment LED systems, and more. We are also going to dig into Max 6, and especially how we can use Arduino to control some OpenGL visuals on the computer. We'll discover network protocols and how to use Arduino even without any network cables, with Wi-Fi. At last, we'll design a small library together and check some nice tips and tricks to improve our C code.

10
Some Advanced Techniques

In this chapter, we are going to learn different techniques that can be used either together or independently. Each technique developed here is a new tool for your future or current projects. We are going to use EEPROMs to provide Arduino boards with a small memory system that is readable and writable.

We are also going to test communications between the Arduino boards themselves, use GPS modules, make our boards autonomous, and more.

Data storage with EEPROMs

Until now, we learned and used the Arduino boards as totally electricity dependent devices. Indeed, they need current in order to execute tasks compiled in our firmware.

As we noticed, when we switch them off, every living variable and data is lost. Fortunately, the firmware isn't.

Three native pools of memory on the Arduino boards

The Arduino boards based on the ATmega168 chipset own three different pools of memory:

- Flash memory
- SRAM
- EEPROM

The flash memory is also named program space. This is the place where our firmware is stored.

The **SRAM** stands for **Static Random Access Memory** and is the place where the running firmware stores, reads, and manipulates variables.

The **EEPROM** stands for **Electrically Erasable Programmable Read-Only Memory**. It is the place where we, programmers, can store things for long-term purposes. This is the place where our firmware sits, and anything in the EEPROM isn't erased should the board be switched off.

ATmega168 has:

- 16000 bytes of Flash (2000 bytes are used for the bootloader)
- 1024 bytes of SRAM
- 512 bytes of EEPROM

Here we won't discuss the fact that we have to take care of the memory while programming; we will do that in the last chapter of this book *Chapter 13, Improving your C Programming and Creating Libraries*.

The interesting part here is the EEPROM space. It allows us to store data on the Arduino and we didn't even know that until now. Let's test the EEPROM native library.

Writing and reading with EEPROM core library

Basically, this example doesn't require any wiring. We are going to use the internal EEPROM of 512 bytes. Here is some code that reads all the bytes of the EEPROM and prints it to the computer's Serial Monitor:

```
#include <EEPROM.h>

// start reading from the first byte (address 0) of the EEPROM
int address = 0;
byte value;

void setup()
{
  // initialize serial and wait for port to open:
  Serial.begin(9600);
}

void loop()
{
  // read a byte from the current address of the EEPROM
  value = EEPROM.read(address);
```

```
Serial.print(address);
Serial.print("\t");
Serial.print(value, DEC);
Serial.println();

// advance to the next address of the EEPROM
address = address + 1;

// there are only 512 bytes of EEPROM, from 0 to 511, so if we're
// on address 512, wrap around to address 0
if (address == 512)
  address = 0;

delay(500);
}
```

This code is in the public domain and provided as an example for the EEPROM library. You can find it in your examples folder in the **File** menu of the Arduino IDE, under the folder **Examples | EEPROM**.

At first, we include the library itself. Then we define a variable for storing the current read address. We initialize it at 0, the beginning of the memory register. We also define a variable as a byte type.

In the setup() function, we initialize the serial communication. In loop(), we read the byte at the current address and store it in the variable value. Then we print the result to the serial port. Notice the \t value in the second Serial.print() statement. This stands for tabulation (as in the *Tab* key on a computer keyboard). This writes tabulation to the serial port between the current address printed and the value itself in order to make things more readable.

We advance to the next address. We check if the address equals 512, if that is the case, we restart the address counter to 0 and so on.

We add a small delay. We can write bytes in the same way using EEPROM. write(addr, val); where addr is the address where you want to write the value val.

Be careful, these are bytes (8 bits = 256 possible values). Read and write operations are quite easy on the internal EEPROM, so let's see how it goes with external EEPROMs wired by an I2C connection.

External EEPROM wiring

There are a lot of cheap EEPROM components available in electronics markets. We are going to use the classic 24LC256, an EEPROM implementing I2C for read/write operations and providing 256 kilobits (32 kilobytes) of memory space.

You can find it at Sparkfun: `https://www.sparkfun.com/products/525`. Here is how we can wire its bigger cousin 24LC1025 (1024k bytes) using I2C:

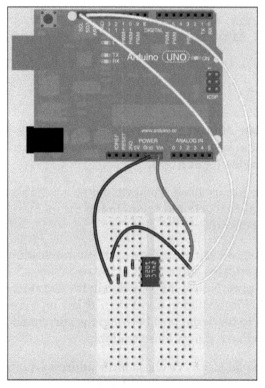

A 24LC256 EEPROM wired to the Arduino via I2C communication

The corresponding diagram is the one shown as follows:

A 24LC256 EEPROM wired to the Arduino via I2C communication

Let's describe the EEPROM.

A0, **A1**, and **A2** are chip address inputs. **+V** and **0V** are **5V** and ground. WP is the write protect pin. If it is wired to ground, we can write to the EEPROM. If it is wired to 5V, we cannot.

SCL and SDA are the two wires involved in the I2C communication and are wired to **SDA / SCL**. **SDA** stands for **Serial Data Line** and **SCL** stands for **Serial Clock Line**. Be careful about the SDA/SCL pins. The following depends on your board:

- The Arduino UNO before R3 and Ethernet's I2C pins are A4 (SDA) and A5 (SCL)
- Mega2560, pins 20 (SDA) and 21 (SCL)
- Leonardo, pin 2 (SDA) and pin 3 (SCL)
- Due Pins, pins 20 (SDA) and 21 (SCL) and also another one SDA1 and SCL1

Reading and writing to the EEPROM

The underlying library that we can use for I2C purposes is `Wire`. You can find it directly in the Arduino core. This library takes care of the raw bits, but we have to look at it more closely.

The `Wire` library takes care of many things for us. Let's check the code in the folder `Chapter10/readWriteI2C`:

```
#include <Wire.h>

void eepromWrite(byte address, byte source_addr, byte data) {
  Wire.beginTransmission(address);
  Wire.write(source_addr);
  Wire.write(data);
  Wire.endTransmission();
}

byte eepromRead(int address, int source_addr) {
  Wire.beginTransmission(address);
  Wire.write(source_addr);
  Wire.endTransmission();

  Wire.requestFrom(address, 1);
  if(Wire.available())
    return Wire.read();
  else
    return 0xFF;
}

void setup() {
  Wire.begin();
  Serial.begin(9600);

  for(int i = 0; i < 10; i++) {
    eepromWrite(B01010000, i, 'a'+i);
    delay(100);
  }

  Serial.println("Bytes written to external EEPROM !");
}

void loop() {
  for(int i = 0; i < 10; i++) {
    byte val = eepromRead(B01010000, i);
```

```
    Serial.print(i);
    Serial.print("\t");
    Serial.print(val);
    Serial.print("\n");
    delay(1000);
  }
}
```

We include the `Wire` library at first. Then we define 2 functions:

- `eepromWrite()`
- `eepromRead()`

These functions write and read bytes to and from the external EEPROM using the `Wire` library.

The `Setup()` function instantiates the `Wire` and the `Serial` communication. Then using a `for` loop, we write data to a specific address. This data is basically a character 'a' plus a number. This structure writes characters from a to a + 9 which means 'j'. This is an example to show how we can store things quickly, but of course we could have written more meaningful data.

We then print a message to the Serial Monitor in order to tell the user that Arduino has finished writing to the EEPROM.

In the `loop()` function, we then read the EEPROM. It is quite similar to the EEPROM library.

Obviously, we still haven't spoken about addresses. Here is an I2C message format:

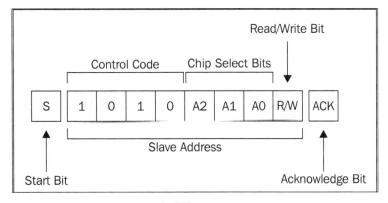

An I2C message

`Wire` library takes care of **Start Bit** and **Acknowledge Bit**. The control code is fixed and you can change the **Chip Select Bits** by wiring **A0**, **A1**, and **A2** pins to ground or +V. That means there are 8 possibilities of addresses from 0 to 7.

1010000 1010001… until 1010111. 1010000 binary means 0x50 in hexadecimal, and 1010111 means 0x57.

In our case, we wired **A0**, **A1**, and **A2** to ground, then the EEPROM address on the I2C bus is 0x50. We could use more than one on the I2C bus, but only if we need more storage capacity. Indeed, we would have to address the different devices inside our firmware.

We could now imagine storing many things on that EEPROM space, from samples for playing PCM audio to, eventually, huge lookup tables or whatever requiring more memory than available on Arduino itself.

Using GPS modules

GPS stands for **Global Positioning System**. This system is based on satellite constellations.

Basically, a receiver that receives signals from at least 4 satellites embedded with a special atomic clock can, by calculating propagation time of these signals between them and itself, calculate precisely its tri-dimensional position. That sounds magical; it is just trigonometric.

We won't get into the details of this process; instead focus on the parsing of data coming from the GPS modules. You can get more information from Wikipedia: `http://en.wikipedia.org/wiki/Global_Positioning_System`.

Wiring the Parallax GPS receiver module

The Parallax GPS Receiver is based on the PMB-248 specification and provides a very easy way to add position detection to the Arduino with its small footprint and low cost.

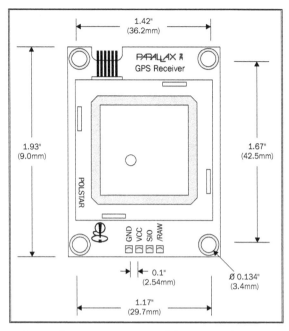

The Parallax GPS Receiver: small footprint and accurate

It provides standard raw NMEA01823 strings or even specific user-requested data via the serial command interface. It can track 12 satellites and even WAAS (system only available in USA and Hawaii for helping the GPS Signal calculation).

NMEA0183 is a combined hardware and logic specification for communication between marine electronic devices such as sonars, anemometers, and many others including GPS. A great description of this protocol can be found here: `http://aprs.gids.nl/nmea/`.

The module provides current time, date, latitude, longitude, altitude speed, and travel direction/heading, among other data.

We can write data to the GPS modules in order to request specific strings. However, if we pull the **/RAW** pin low, some strings are automatically transmitted by the modules. These strings are:

- $GPGGA: Global Positioning System Fix Data
- $GPGSV: GPS satellites in view
- $GPGSA: GPS DOP and active satellites
- $GPRMC: Recommended minimum specific GPS/Transit data

This data has to be grabbed by the Arduino and eventually used. Let's check the wiring first:

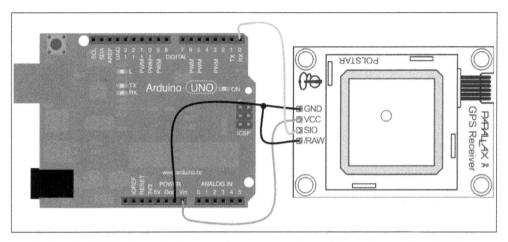

The Parallax GPS Receiver wired to the Arduino in automatic mode by pulling low the /RAW pin

The wiring is quite easy.

Yes, the Parallax GPS Receiver only consumes one data pin: digital pin 0. Let's stop here for two seconds. Didn't we talk about the fact that we cannot use the USB port for serial monitoring and pins 0 and 1 for other serial feature at the same time on Arduino?

With serial communication using Rx/Tx 2 wires, serial software implementation can be full duplex.

In our case, the GPS device sends data to the Arduino Rx pin. This pin (digital pin 0) is wired to the USB Rx pin. At the same time, the Arduino sends data to the computer using the USB Tx pin that is wired to digital pin 1.

Is there a problem in our case here? No. We just have to take care of interferences. We must not send data from the computer via USB to the Arduino because it already receives data on its serial pin 0 from the GPS device. This is the only thing we have to be careful of.

The `Serial.write()` function will write to digital pin 1, and the USB Tx digital pin 1 isn't wired to anything. Therefore, no problem, data will be sent to the USB. The `Serial.read()` function reads from digital pin 0 and USB and we don't send anything from the computer to USB, so it can read digital pin 0 without any problem.

We pull the /RAW pin to low. In this mode, the device pops data out to the Arduino automatically; I mean, without having to request it.

Parsing GPS location data

Before building any firmware that will be able to use GPS data, we have to know a bit more about what the device is able to transmit.

We can read the datasheet of the GPS device at: `http://www.rcc.ryerson.ca/medi` `a/2008HCLParallaxGPSReceiverModuledatasheet.pdf`.

Here is an example of data that can be transmitted:

`$GPRMC,220516,A,5133.82,N,00042.24,W,173.8,231.8,130694,004.2,W*70`

`$GPRMC` defines the type of information sequence sent. The comma is a separator that separates each data field.

Here is the meaning of each field:

1. UTC time of fix
2. Data status (`A` means valid position and `V` means warning)
3. Latitude of the fix
4. North or South latitude
5. Longitude of the fix
6. East or West longitude
7. Speed over ground (in knots)
8. Track made good in degrees
9. UTC date of fix
10. Magnetic variation in degrees
11. East or West magnetic variation
12. Checksum

As soon as we know what data is sent, we can code a parser in our firmware. Here is a possible firmware. You can find it in folder `Chapter10/locateMe`:

```
int rxPin = 0;                      // RX PIN
int byteGPS = -1;                   // Current read byte
char line[300] = "";                // Buffer
char commandGPR[7] = "$GPRMC";      // String related to messages

int counter=0;
int correctness=0;
int lineCounter=0;
int index[13];
```

```
void setup() {

  pinMode(rxPin, INPUT);
  Serial.begin(4800);

  // Clear buffer
  for (int i=0;i<300;i++){
    line[i]=' ';
  }
}


void loop() {

  byteGPS = Serial.read();              // Read a byte from the serial port

  // Test if the port is empty
  if (byteGPS == -1) {
    delay(100);
  }

  else {  // if it isn't empty

    line[lineCounter] = byteGPS;    // put data read in the buffer
    lineCounter++;

    Serial.print(byteGPS);    // print data read to the serial monitor

    // Test if the transmission is finished
    // if it is finished, we begin to parse !
    if (byteGPS==13){

      counter=0;
      correctness=0;

      // Test if the received command starts by $GPR
      // If it does, increase correctness counter
      for (int i=1;i<7;i++){
        if (line[i]==commandGPR[i-1]){
          correctness++;
        }
      }


      if(correctness==6){
        // We are sure command is okay here.

        //
        for (int i=0;i<300;i++){
```

```
  // store position of "," separators
  if (line[i]==','){
    index[counter]=i;
    counter++;
  }

  // store position of "*" separator meaning the last byte
  if (line[i]=='*'){     // ... and the "*"
    index[12]=i;
    counter++;
  }
}

// Write data to serial monitor on the computer
Serial.println("");
Serial.println("");
Serial.println("--------------");
for (int i=0;i<12;i++){
  switch(i){
  case 0 :
    Serial.print("Time in UTC (HhMmSs): ");
    break;
  case 1 :
    Serial.print("Status (A=OK,V=KO): ");
    break;
  case 2 :
    Serial.print("Latitude: ");
    break;
  case 3 :
    Serial.print("Direction (N/S): ");
    break;
  case 4 :
    Serial.print("Longitude: ");
    break;
  case 5 :
    Serial.print("Direction (E/W): ");
    break;
  case 6 :
    Serial.print("Velocity in knots: ");
    break;
  case 7 :
    Serial.print("Heading in degrees: ");
    break;
  case 8 :
    Serial.print("Date UTC (DdMmAa): ");
    break;
  case 9 :
    Serial.print("Magnetic degrees: ");
    break;
```

```
                case 10 :
                   Serial.print("(E/W): ");
                   break;
                case 11 :
                   Serial.print("Mode: ");
                   break;
                case 12 :
                   Serial.print("Checksum: ");
                   break;
                }
                for (int j=index[i];j<(index[i+1]-1);j++){
                   Serial.print(line[j+1]);
                }
                Serial.println("");
             }
             Serial.println("--------------");
          }

          // Reset the buffer
          lineCounter=0;
          for (int i=0;i<300;i++){
             line[i]=' ';
          }
       }
     }
  }
```

Let's explain the code a bit. At first, I'm defining several variables:

- rxPin is the digital input where the GPS device is wired
- byteGPS is the latest byte read from the GPS using serial communication
- line is a buffer array
- commandGPR is a string related to messages we want to parse
- counter is the index of the index array
- correctness stores the validity of the message
- lineCounter is a counter keeping track of the buffer position of the data
- index stores the position of each of the separators in the GPS data string (",")

In the `setup()` function, we first define digital pin 0 as an input, and then start the serial communication with a rate of 4800 baud as required by serial interface of the Parallax GPS Receiver (remember to always check your date sheets). Then, we are clearing our `line` array buffer by filling it with a space character.

In the `loop()` function, we begin by reading byte from serial input, the digital pin being 0. If the port isn't empty, we enter it in the second part of the `if` conditional test defined by the `else` block. If it is empty, we just wait for 100 ms then try to read it again.

At first, the parsing begins by putting the data read in the line buffer at this particular index of the array: `lineCounter`. Then, we increment the latter in order to store the data received.

We then print the data read as a raw line to the USB port. It is at this moment that the Serial Monitor can receive and display it as the raw data row we quoted before as an example.

Then, we test the data itself, comparing it to 13. If it equals 13, it means data communication is finished and we can begin to parse.

We reset the `counter` and `correctness` variables and check if the first 6 characters in the buffer equals $GPRMC. For each match, we increment the `correctness` variable.

This is a classic pattern. Indeed, if all the tests are true, it means `correctness` equals 6 at the end. Then we just have to check if `correctness` equals 6 to see if all the tests have been true, and if the first 6 characters equals $GPRMC.

If this is the case, we can be sure we have a correct NMEA raw sequence of the type $GPRMC, and we can start to actually parse the payload part of the data.

At first, we split our raw string by storing the position in the string of each comma separator. We then do the same with the last part separator, the "*" character. At this point, we are able to distinguish which character belongs to which part of the string, I mean, which part of the raw message.

It is a loop between each value of the raw message, and we test each value using a switch/case structure in order to display the correct sentence introducing each value of the GPS Data message.

The most tricky part, finally, is the last `for()` loop. We don't start as usual. Indeed, we start the j index in the loop using the array `index` at the specific position i.

Here is a small schematic showing indexes around the raw message:

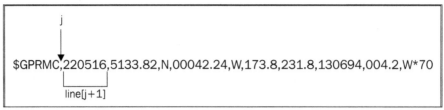

Progressively parsing each part of the message according to each separator

We increment progressively according to each separator's position, and we display each value. This is one way of parsing and using location data using a GPS module. This data can be used in many ways, depending on your purpose. I like data visualization, and I made small projects for students with a GPS module grabbing location every 30s while walking in the street and writing it on an EEPROM. Then, I used this data to make some graphs. One I liked a lot is the following:

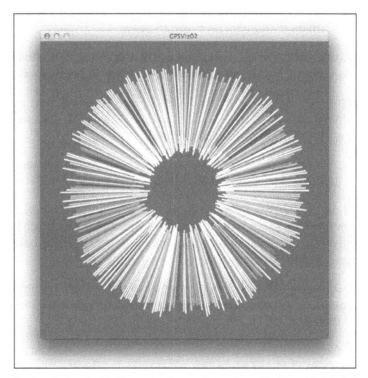

Data visualization designed with Processing from a data set provided by a GPS Arduino module

Each line is a timestamp. The size of the line represents the time we spent between two measures of my Arduino-based GPS module. The longer the line is, the more time I spent at this step of the travel.

Your question could be: How did you supply power to your Arduino + GPS module walking in the street?

Now, let's check how we can make the Arduino autonomous using batteries.

Arduino, battery, and autonomy

The Arduino boards can supply power in two ways:

- A USB wire from a computer
- An external power supply

We already used USB for supplying power to the Arduino since the beginning of the section. This is a pretty nice way to begin (and even to make a great project). This is easy and works for many purposes.

We can also use an external power supply when we need more autonomy and mobility with our Arduino devices.

In any case, we have to keep in mind that both our Arduino and our circuits wired to it need power. Usually, the Arduino consumption is no more than 50mA. Add some LEDs and you'll see the consumption increase.

Let's check some cases of real use.

Classic cases of USB power supplying

Why and when would we use a USB power supply?

Obviously, if we need our computer connected to our Arduino for data communication purposes, we can naturally supply power to the Arduino through the USB.

This is the main reason for using a USB power supply.

There are also some cases where we cannot have a lot of power sockets. Sometimes, there are many constraints in installation design projects and we don't have a lot of power sockets. This is also one case of supplying power using the USB.

Basically, the first thing to bear in mind before using power supplied by the USB port is the global consumption amount of our circuit.

Indeed, as we have already learned, the maximum current a USB port can provide is around 500mA. Be sure you don't exceed this value. Above this limit of consumption, things become totally unpredictable and some computers can even reboot while some others can disable all USB ports. We have to keep that in mind.

Supplying external power

There are two different ways to supply power to an Arduino-based system. We can state the two main power supplies as:

- Batteries
- Power adapters

Supplying with batteries

If we remember correctly, the Arduino Uno and Mega for instance can operate on an external power supply of 6 V to 20 V. For stable use, the recommended range is 7 V to 12 V. 9 V is an ideal voltage.

In order to set the board to external power supply, you have to take care of the power jumper. We have to put it on the external power supply side, named EXT. This setup is for the Arduino Diecimilla and older the Arduino boards:

The power supply jumper put on the EXT side, meaning set up to external power supply

Let's check the basic wiring with a 9 V battery:

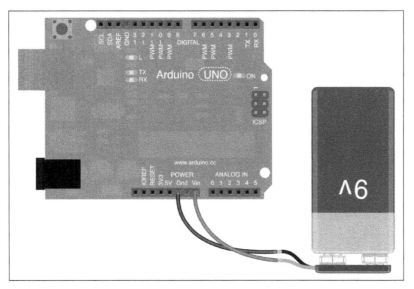

A 9V battery wired to an Arduino board UNO R3

This simple wiring provides a way to supply power to the Arduino board. If you plug some other circuits to the Arduino, the battery through the Arduino will feed them.

There are also some other types of batteries that we can use. Coin cell batteries are a nice way to save space while supplying power externally:

A classic coin cell battery

There are many type of coin cell holders to use this type of battery in our circuits. Usually, coin cell batteries provide 3.6 V at 110 mAh. If this cannot supply power to the Arduino Uno, it can easily supply the Arduino Pro Mini working at a voltage of 3.3 V:

Arduino Pro Mono

The Arduino Pro Mini board is really interesting as it can be embedded in many circuits that need to be discrete and sometimes hidden in walls for digital art installations or put into a small plastic box that can be carried in a pocket when they are used as a mobile tool.

We can also use polymer lithium-ion batteries. I used them a couple of times for an autonomous device project.

However, we can have some projects that require more power.

Power adapter for Arduino supply

For projects requiring more power, we have to use an external power supply. The setup of the Arduino stays the same as with batteries. The off-the-shelf Arduino adapter has to meet some requirements:

- DC adapter (No AC adapter here!)
- Output voltage between 9V and 12V DC
- Able to output a minimum current of 250mA but aim at 500mA at least or preferably 1A
- Must have a centre positive 2.1mm power plug

Here are the patterns you have to look for on your adapter before plugging in the Arduino.

First, the center of the connector has to be the positive part; check out the following diagram. You should see that on an Arduino-compatible adapter:

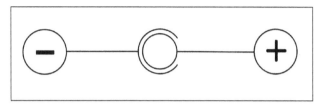

The symbol showing the center positive plug

Then, the voltage and current characteristics. This has to show something like: **OUTPUT: 12 VDC 1 A**. This is an example; 12 VDC and 5 A is also fine. Don't forget that current is only driven by what's there in your circuit. A power adapter that puts out a higher current will not harm your circuit, because a circuit will only draw what it needs.

A lot of adapters are available in the market and can be used with our Arduino boards.

How to calculate current consumption

In order to calculate current in your circuit, you have to use Ohm's law as described in the first chapter of this book.

When you check the datasheet of a component, like an LED, you can see that the current passed through it.

Let's check the RGB LED Common Cathode with this datasheet: `https://www.sparkfun.com/datasheets/Components/YSL-R596CR3G4B5C-C10.pdf`

We can see a forward current of 20 mA and a peak forward current of 30 mA. If we have five LEDs like that switched on at the maximum brightness (that is red, blue, and green lighted up), we have: 5 x (20 + 20 + 20) = 300 mA needed for normal use and even peaks would consume 5 x (30 + 30 + 30) = 450 mA.

This is in the case where all LEDs are fully switched on at the same time.

You must have understood the strategy we already used in power supply cycling, switching on each LED one after the other in quick succession. This provides a way to reduce the power consumption and also allow some projects to use a lot of LEDs without requiring an external power adapter.

I won't describe the calculations for each case here, but you'd have to refer to electricity rules to precisely calculate the consumption.

By experience, there is nothing better than your voltmeter and Ampere meter, the former measuring voltage between two points and the latter measuring current at some points along the circuit.

I'd suggest that you make some calculations to be sure to:

- Not override the Arduino capacity per pins
- Not override USB 450mA limit, in case you use a USB power supply

Then, after that, begin to wire and measure at the same time with voltmeter and Ampere meter.

At last, a classic reference for most of the Arduino boards is available at this page: `http://playground.arduino.cc/Main/ArduinoPinCurrentLimitations`.

We can find the limitations for current consumption for each part of the Arduino.

Drawing on gLCDs

Drawing is always fun. Drawing and handling LCD displays instead of LEDs matrices is really interesting too, because we have devices with high-density points we can switch on and off easily.

LCDs exist in many types. The two main types are the character and graphical type.

We are talking about the graphical type here, especially those based on the KS0108 graphics-only controller used in many regular gLCD devices.

We are going to use a nice library that is available on Google. It has code by Michael Margolis and Bill Perry, and it is named `glcd-arduino`. This library is licensed under the GNU Lesser GPL.

Let's download it here: `http://code.google.com/p/glcd-arduino/downloads/list`. Download the most recent version.

Unzip it, put it in the place where all your libraries are, and restart or start your Arduino IDE.

You should now see a lot of examples related to the gLCD library.

We won't check all the nice features and functions provided by this library here, but you can check this page on the Arduino website: `http://playground.arduino.cc/Code/GLCDks0108`.

Wiring the device

We are going to check the wiring of a KS0108 based gLCD type Panel B:

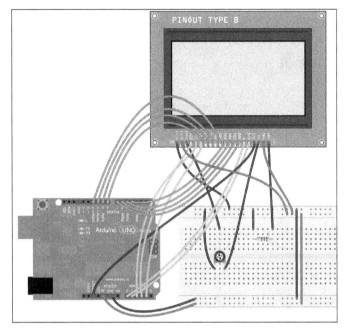

A lot of wires wiring the gLCD to Arduino and the potentiometer to adjust LCD contrast

The corresponding electrical diagram is as follows:

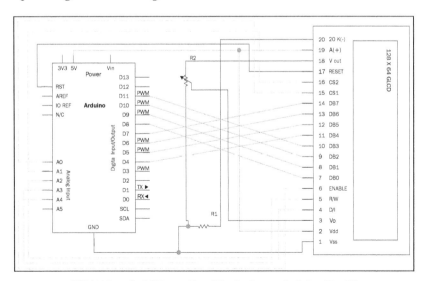

KS0108 based gLCD type Panel B wired to an Arduino Uno R3

These are a lot of wires. Of course, we can multiply things. We can also use an Arduino MEGA and keep using the other digital pin available for other purposes, but that is not the point here. Let's check some of the functions of this powerful library.

Demoing the library

Take the example named GLCDdemo. It shows you almost all the functions available in the library.

There is very good PDF documentation provided with the library. It explains each available method. You can find it in the library folder in the doc subfolder:

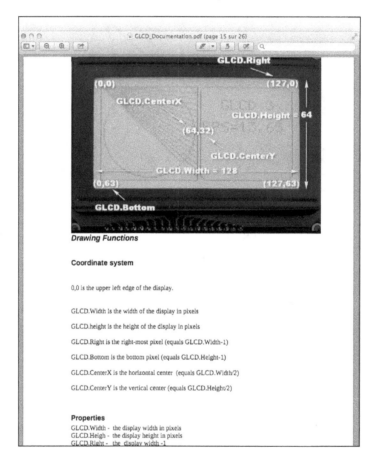

The documentation of gLCD-Arduino showing the screen coordinates system

At first, we have to include `glcd.h` in order to use the library. Then, we have to include some other headers, in this example, fonts and bitmap in order to use the font typographic methods and the bitmap objects too.

Some useful method's families

I'd suggest ordering learning methods into three parts:

- Global GLCD methods
- Drawing methods
- Text methods

Global GLCD methods

The first is the `init()` function. This one initializes the library and has to be called before any other gLCD methods.

The `SetDisplayMode()` function is useful because it sets up the use of the LCD as normal (writing in black over white background) or inverted. White just means not black. The real color depends on the backlight color, of course.

The `ClearScreen()` function erases the screen, filling it with white background in normal mode, or black in inverted mode.

The `ReadData()` and `WriteData()` functions are really raw methods that get and set the byte of data at particular coordinates.

Drawing methods

These are a set of functions dedicated to drawing on the screen.

The set of constants are as follows:

- `GLCD.Width` is the display width in pixels
- `GLCD.Height` is the display height in pixels
- `GLCD.Right` is the last pixel column at the right (equals GLCD.Width – 1)
- `GLCD.Bottom` is the last pixel row at the bottom (equals GLCD. Height – 1)
- `GLCD.CenterX` and `GLCD.CenterY` are the coordinates of the pixel in the middle

Basically, you can draw by moving the graphics cursor and by drawing primitive shapes:

Function	Description
GotoXY()	Moves the cursor to specific coordinates
DrawVLine()	Draws a vertical line from a point to another point in the same pixel column but above or below the initial point
DrawHLine()	Works the same as DrawVLine() but on the same pixel row
DrawLine()	Draws a line between two coordinates

Some other, more complex shapes can be drawn too:

Functions	Descriptions
DrawRect()	Draws a rectangle from a point when provided with a width and height.
FillRect()	Works the same as DrawRect(), but by filling the rectangle shape with black (or white) pixels.
DrawRoundRect()	Draws a nice rectangle with rounded corners.
DrawCircle() and FillCircle()	Draws a circle from coordinates and a radius, and a circle filled with black (or white) pixels.
DrawBitmap()	Draws a whole bitmap at a particular position on the screen. It uses a pointer to that bitmap in memory.

With this set of functions, you can basically draw anything you want.

Text methods

These are a set of functions dedicated to typography on the screen:

Functions	Descriptions
SelectFont()	At first, this chooses the font to be used in the next functions calls.
SetFontColor()	Chooses the color.
SetTextMode()	Chooses a scrolling direction.
CursorTo()	Moves the cursor to a specific column and row. The column calculation uses the width of the widest character.
CursorToXY()	Moves the cursor to particular pixel coordinate.

One important feature to know about, is the fact that Arduino's print functions can be used with gLCD library; `GLCD.print()` works fine, for instance. There are also a couple of other functions available that can be found on the official website.

At last, I'd suggest you to test the example named `life`. This is based on the John Conway's Game of Life. This is a nice example of what you can do and implement some nice and useful logic.

Drawing on gLCD is nice, but we could also use a small module handling VGA.

Using VGA with the Gameduino Shield

Gameduino is an Arduino Shield. This is the first one we are using here in this book. Basically, a shield is a PCB (printed circuit board) that can be plugged to another PCB, here our Arduino.

Arduino Shields are pre-made circuits including components and sometimes processors too. They add features to our Arduino board by handling some specific tasks.

Here, the Gameduino will add VGA drawing abilities to our Arduino that can't be done on its own.

The Gameduino adds a VGA port, a mini-jack for the sound, and also includes an FPGA Xilling Spartan3A. FPGA Xilling Spartan3A can process graphical data faster than the Arduino itself. Arduino can control this graphical hardware driver by SPI interface.

Let's see how it works:

The Gameduino controller Arduino Shield

Arduino Shields can be plugged in Arduino boards directly. Check the following screenshot:

The Gameduino plugged in the Arduino board

Here are some characteristics of the Gameduino:

- Video output is 400 x 300 pixels in 512 colors
- All color processed internally at 15 bit precision
- Compatible with any standard VGA monitor (800 x 600 @ 72 Hz)
- Background graphics (512 x 512 pixel character, 256 characters)
- Foreground graphics (sprite 16 x 16 abilities, transparency, rotate/flip, and sprite collision detection)
- Audio output as stereo; 12-bit frequency synthesizer
- 64 independent voices at 10 to 8000 hz
- Sample playback channel

The underlying concept is to plug it in the Arduino and to control it using our Arduino firmware with the library taking care of all SPI communication between the Arduino and Gameduino.

We cannot describe all the examples right here in this book, but I want to point you in the right direction. At first, the official website: `http://excamera.com/sphinx/gameduino/`.

You can find the library here: `http://excamera.com/files/gameduino/synth/sketches/Gameduino.zip`.

You can also check and use the quick reference poster here: `http://excamera.com/files/gameduino/synth/doc/gen/poster.pdf`.

For your information, I'm currently designing a piece on digital art installation based on this shield. I intend to describe it on my own website `http://julienbayle.net` and the whole schematics will be provided too.

Summary

In this first, advanced chapter, we learned a bit more about how to deal with new concrete concepts such as storing data on non-volatile memories (internal and external EEPROM), use GPS module receivers, draw on graphical LCD, and use a nice Arduino Shield named Gameduino to add new features and power to our Arduino. This allowed it to display a VGA signal and also to produce audio. We also learned the use of Arduino as a very portable and mobile device, autonomous from the power supply point of view.

In the next chapter, we are going to talk about networking concepts. Creating and using networks are usual ways of communication today. We will describe wired and wireless network use with our Arduino projects in the next chapter.

11
Networking

In this chapter we are going to talk about linking objects and making them talk by creating communication networks. We are going to learn how we can make multiple Arduinos and computers communicate using network links and protocols.

After defining what a network is (specifically, a data network), we'll describe ways to use wired Ethernet links between Arduinos and computers. This will open the Arduino world to the Internet. Then, we'll discover how to create Bluetooth communications.

We will learn how to use Ethernet Wi-Fi in order to connect the Arduino to computers or other Arduinos without being tethered by network cables.

At last, we will study a couple of examples from the one in which we will fire message to the micro-blogging service Twitter, to the one in which we will parse and react to data received from the Internet.

We will also introduce the OSC exchange protocol, widely used in anything related to interaction design, music, and multimedia.

An overview of networks

A network is a system of elements linked together. There are many networks around us such as highway systems, electrical grids, and data networks. Data networks surround us. They relate to video services networks, phone and global telecommunication networks, computer networks, and so on. We are going to focus on these types of networks by talking about how we can share data over different types of media such as wires transporting electric pulses or electromagnetic waves facilitating wireless communication.

Before we dive into the details of network implementations for Arduino boards, we are going to describe a model named the OSI model. It is a very useful representation of what a data network is and what it involves.

Overview of the OSI model

The **Open Systems Interconnection** model (**OSI** model) has been initiated in 1977 by the International Organization for Standardization in order to define prescriptions and requirements around the functions of communication systems in terms of abstract layers.

Basically, this is a layers-based model describing what features are necessary to design communicating systems. Here is the OSI model with seven layers:

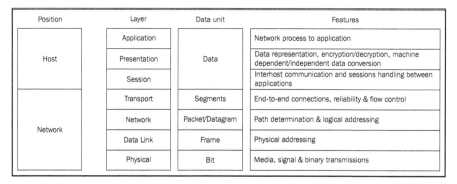

Position	Layer	Data unit	Features
Host	Application	Data	Network process to application
	Presentation		Data representation, encryption/decryption, machine dependent/independent data conversion
	Session		Interhost communication and sessions handling between applications
Network	Transport	Segments	End-to-end connections, reliability & flow control
	Network	Packet/Datagram	Path determination & logical addressing
	Data Link	Frame	Physical addressing
	Physical	Bit	Media, signal & binary transmissions

OSI model describing communication system requirements with seven abstraction layers

Protocols and communications

A communications protocol is a set of message formats and rules providing a way of communication between at least two participants. Within each layer, one or more entities implements its functionality and each entity interacts directly and only with the layer just beneath it and at the same time provides facilities for use by the layer above it. A protocol enables an entity in one host to interact with a corresponding entity at the same layer in another host. This can be represented by the following diagram:

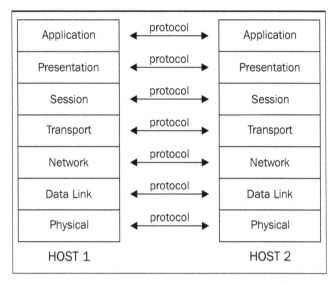

Protocols helping hosts' layers to communicate together

Data encapsulation and decapsulation

If the application of one host needs to send data to another application of another host, the effective data, also named the payload, is passed down directly to the layer beneath it. In order to make the application able to retrieve its data, a header and footer are added to this data depending on the protocol used at each layer. This is called **encapsulation** and it happens until the lowest layer, which is the physical one. At this point, a flow of bits is modulated on the medium for the receiver.

The receiver has to make the data progressively climb the layer stack, passing data from a layer to a higher layer and addressing it to the right entities in each layer using previously added headers and footers. These headers and footers are removed all along the path; this is called **decapsulation**.

At the end of the journey, the application of the receiver receives its data and can process it. This whole process can be represented by the following diagram:

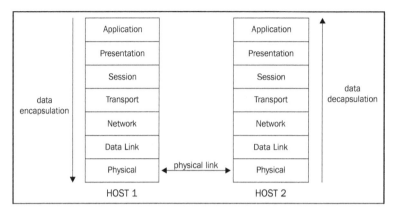

Encapsulation and decapsulation all along the layers' stack

We can also represent these processes as shown in the following figure. The small gray rectangle is the data payload for the layer N+1.

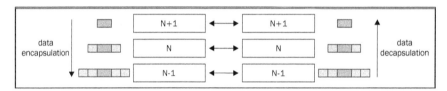

Adding and removing specific headers and footers according to the protocols used

At each level, two hosts interact using a protocol transmitted, which we call **Protocol Data Unit** or **PDU**. We also call **Service Data Unit** or **SDU**, a specific unit of data passed down from a layer to a lower layer and that has not yet been encapsulated.

Each layer considers the data received as data for it and adds/removes headers and footers according to the protocol used.

We are now going to illustrate each layer and protocol by examples.

The roles of each layer

We are going to describe the purpose and roles of each layer here.

Physical layer

The physical layer defines electrical and physical specifications required for communication.

Pin layout, voltages and line impedance, signal timing, network adapters, or host bus adapters are defined in this layer. Basically, this layer performs three major functions/services:

- Initialization and termination of a connection to a communication medium
- Participation in shared resources control processes
- Conversion between the data communicated and the electrical signals which carry them

We can quote some known standards being in this physical layer:

- ADSL and ISDN (network and phony provider services)
- Bluetooth
- IEEE 1394 (FireWire)
- USB
- IrDA (data transfer over infrared links)
- SONET, SDH (wide area optic fiber networks operated by providers)

Data link layer

This layer is made of two sublayers:

- Logical Link Control (LLC)
- Media Access Control (MAC)

Both are responsible for transferring data between network entities and to detect errors that may occur at the physical layer, and eventually to fix them. Basically, this layer provides these functions/services:

- Framing
- Physical addressing
- Flow control
- Error control
- Access control
- Media access control

We can quote some known standards of this data link layer:

- Ethernet
- Wi-Fi
- PPP
- I2C

We have to keep in mind that the second layer is also the domain of local area networks with only physical addresses. It can be federated using LAN switches.

By the way, we often need to segment networks and also communicate wider and so we need another addressing concept; this introduces the network layer.

Network layer

This layer provides the way to transfer data sequences between hosts that can be in different networks. It provides the following functions/services:

- Routing
- Fragmentation and reassembly
- Delivery error reports

Routing provides a way to make hosts on a different network able to communicate by using a network addressing system.

Fragmentation and reassembly also occur at this level. These provide a way to chop data streams into pieces and to be able to reassemble parts after the transmission. We can quote some known standards in this layer:

- ARP (resolving and translating physical MAC address into network address)
- BOOTP (providing a way for the host to boot over the network)
- BGP, OSPF, RIP, and other routing protocols
- IPv4 and IPv6 (Internet Protocol)

Routers are usually the gear where the routing occurs. They are connected to more than one network and make data going from one network to another. This is also the place where we can put some access lists in order to control access based on IP addresses.

Transport layer

This layer is in charge of the data transfer between end users, being at the crossroads of network layers and application layers. This layer provides the following functions/services:

- Flow control to assure reliability of the link used
- Segmentation/desegmentation of data units
- Error control

Usually, we order protocols in two categories:

- State-oriented
- Connection-oriented

This means this layer can keep track of segments emitted and eventually retransmit them in case of previously failed transmission.

We can quote the two well-known standards of the IP suite in this layer:

- TCP
- UDP

TCP is the connection-oriented one. It keeps the communication reliable by checking a lot of elements at each transmission or at each x segments transmitted.

UDP is simpler and stateless. It doesn't provide a communication state control and thus is lighter. It is more suited for transaction-oriented query/response protocol such as DNS (Domain Name System) or NTP (Network Time Protocol). If there is something wrong, such as a segment not transmitted well, the above layer has to take care of resending a request, for instance.

Application/Host layers

I grouped the highest three layers under the terms application and host.

Indeed, they aren't considered as network layers, but they are part of OSI model because they are often the final purpose of any network communication.

We find a lot of client/server applications there:

- FTP for basic and light file transfers
- POP3, IMAP, and SMTP for mail services
- SSH for secure remote shell communication
- HTTP for web server browsing and downloading (and nowadays much more)

We also find a lot of standards related to encryption and security such as TLS (Transport Layer Security). Our firmware, an executing Processing code, Max 6 running a patch are in this layer.

If we want to make them communicate through a wide variety of networks, we need some OSI stack. I mean, we need a transport and network protocol and a medium to transport our data.

If our modern computers own the whole network stack ready to use, we have to build this later in our Arduino's firmware if we want them to be able to communicate with the world. This is what we are going to do in the next subchapter.

Some aspects of IP addresses and ports

One of the protocol stacks we tend to use each day is the TCP/IP one. TCP is the layer 4 transport protocol, and IP the layer 3 network.

This is the most used network protocol in the world both for end users and for companies.

We are going to explain a little bit more about the IP addressing system, subnet masks, and communication ports. I won't be describing a complete network course.

The IP address

An IP address is a numerical address referenced by any devices wanting to communicate over an IP network. IP is currently used in 2 versions: IPv4 and IPv6. We are considering IPv4 here because it is currently the only one used by end users. IPv4 addresses are coded over 32 bits. They are often written as a human-readable set of 4 bytes separated by a point. 192.168.1.222 is the current IP address of my computer. There are 2^{32} possible unique addresses and all aren't routable over the Internet. Some are reserved for private use. Some companies assign Internet-routable addresses. Indeed, we cannot use both addresses as this is handled by global organizations. Each country has sets of addresses attributed for their own purposes.

The subnet

A subnet is a way to segment our network into multiple smaller ones. A device network's configuration contains usually the address, the subnet mask, and a gateway.

The address and the subnet mask define the network range. It is necessary to know if a transmitter can communicate directly with a receiver. Indeed, if the latter is inside the same network, communication can occur directly; if it is on another network, the transmitter has to send its data to the gateway that will route data to the correct next node on the networks in order to reach, if possible, the receiver.

The gateway knows about the networks to which it is connected. It can route data across different networks and eventually filter some data according to some rules.

Usually, the subnet mask is written as a human-readable set of 4 bytes too. There is obviously a bit notation, more difficult for those not used to manipulating the numbers.

The subnet mask of my computer is 255.255.255.0. This information and my IP address defines that my home network begins at 192.168.1.0 (which is the base network address) and finish at 192.168.1.255 (which is the broadcast address). I cannot use these addresses for my device, but only those from 192.168.1.1 to 192.168.1.254.

The communication port

A communication port is something defined and related to layer 4, the transport layer.

Imagine you want to address a message to a host for a particular application. The receiver has to be in a listening mode for the message he wants to receive.

This means it has to open and reserve a specific socket for the connection, and that is a communication port. Usually, applications open specific ports for their own purpose, and once a port has been opened and reserved by an application, it cannot be used by another application while it is opened by the first one.

This provides a powerful system for data exchange. Indeed, if we want to send data to a host for more than one application, we can specifically address our messages to this host on a different port to reach different applications.

Of course, standards had to be defined for global communications.

TCP port 80 is used for the HTTP protocol related to data exchange with web-servers.

UDP port 53 is used for anything related to DNS.

If you are curious, you can read the following huge official text file containing all declared and reserved port and the related services: `http://www.ietf.org/assignments/service-names-port-numbers/service-names-port-numbers.txt`.

These are conventions. Someone can easily run a web server on a port other than 80. Then, the specific clients of this web server would have to know about the port used. This is why conventions and standards are useful.

Wiring Arduino to wired Ethernet

Ethernet is the local area network most used nowadays.

Usual Arduino boards don't provide Ethernet ability. There is one board named Arduino Ethernet that provides native Ethernet and network features. By the way, it doesn't provide any USB-native features.

You can find the reference page here: `http://arduino.cc/en/Main/ArduinoBoardEthernet`.

Arduino Ethernet board with the Ethernet connector

We are going to use the Arduino Ethernet Shield and a 100BASE-T cable with the Arduino UNO R3. It keeps the USB features and adds Ethernet network connectivity and provides a nice way to link our computer to the Arduino with a much longer cable that USB ones.

The Arduino Ethernet Shield

If you look for the Arduino Ethernet module, you must know they are sold either with or without the PoE module.

PoE stands for **Power over Ethernet** and is a way to supply power to devices through Ethernet connections. This requires two parts:

- A module on the device that has to be supplied
- A network equipment able to provide PoE support

In our case here, we won't use PoE.

Making Processing and Arduino communicate over Ethernet

Let's design a basic system showing how to set up a communication over Ethernet between the Arduino board and a processing applet.

Here, we are going to use an Arduino board wired to our computer using Ethernet. We push a button that triggers the Arduino to send a message over UDP to the Processing applet on the computer. The applet reacts by drawing something and sends back a message to the Arduino, which switches on its built-in LED.

Basic wiring

Here, we are wiring a switch and using the built-in LED board. We have to connect our Arduino board to our computer using an Ethernet cable.

This wiring is very similar to the MonoSwitch project in *Chapter 5* except that we are wiring the Arduino Ethernet Shield here instead of the Arduino board itself.

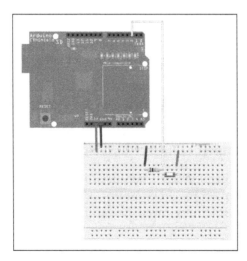

The switch and the pull-down resistor wired to the Arduino Ethernet Shield

The corresponding circuit diagram is as follows:

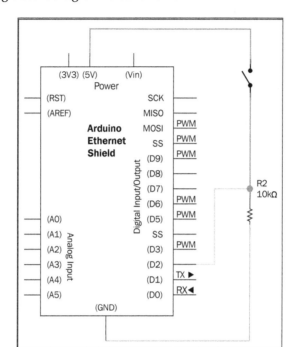

The switch and the pull-down resistor wired to the Arduino Ethernet Shield

Coding network connectivity implementation in Arduino

As we described, if we want to give our Arduino the ability to communicate over the Ethernet cable, and more generally over an Ethernet network, we have to implement the required standards in the firmware.

There is a library called Ethernet that can provide a great number of features.

As usual, we have to include this native library itself. You can choose to do that by navigating to **Sketch | Import Library**, which includes almost everything you need.

However, since Arduino version 0018, because of the implementation of SPI and because the Arduino Ethernet Shield communicates with the Arduino board through SPI, we have to include something more. Be careful about that.

For this code, you need:

```
#include <SPI.h>
#include <Ethernet.h>
#include <EthernetUdp.h>
```

This is a example of the Arduino code, followed by an explanation.

You can find the complete Arduino code at `Chapter11/WiredEthernet`.

```
#include <SPI.h>
#include <Ethernet.h>
#include <EthernetUdp.h>

// Switch & LED stuff
const int switchPin = 2;      // switch pin
const int ledPin =   13;      // built-in LED pin
int switchState = 0;          // storage variable for current switch
state
int lastSwitchState = LOW;
long lastDebounceTime = 0;
long debounceDelay = 50;


// Network related stuff

// a MAC address, an IP address and a port for the Arduino
byte mac[] = {
  0xDE, 0xAD, 0xBE, 0xEF, 0xFE, 0xED };
IPAddress ipArduino(192, 168, 1, 123);
unsigned int ArduinoPort = 9999;

// an IP address and a UDP port for the Computer
// modify these according to your configuration
IPAddress ipComputer(192, 168, 1, 222);
unsigned int ComputerPort = 10000;

// Send/receive buffer
char packetBuffer[UDP_TX_PACKET_MAX_SIZE]; //buffer for incoming
packets

// Instantiate EthernetUDP instance to send/receive packets over UDP
EthernetUDP Udp;


void setup() {
  pinMode(ledPin, OUTPUT);    // the led pin is setup as an output
  pinMode(switchPin, INPUT); // the switch pin is setup as an input

  // start Ethernet and UDP:
```

```
    Ethernet.begin(mac,ipArduino);
    Udp.begin(ArduinoPort);
}

void loop(){

    // if a packet has been received read a packet into packetBufffer
    if (Udp.parsePacket()) Udp.read(packetBuffer,UDP_TX_PACKET_MAX_
SIZE);
    if (packetBuffer == "Light") digitalWrite(ledPin, HIGH);
    else if (packetBuffer == "Dark") digitalWrite(ledPin, LOW);

    // read the state of the digital pin
    int readInput = digitalRead(switchPin);
    if (readInput != lastSwitchState)
    {
      lastDebounceTime = millis();
    }

    if ( (millis() - lastDebounceTime) > debounceDelay )
    {
      switchState = readInput;
    }

    lastSwitchState = readInput;
    if (switchState == HIGH)
    {
      // If switch is pushed, a packet is sent to Processing
      Udp.beginPacket(ipComputer, ComputerPort);
      Udp.write('Pushed');
      Udp.endPacket();
    }
    else
    {
      // If switch is pushed, a packet is sent to Processing
      Udp.beginPacket(ipComputer, ComputerPort);
      Udp.write('Released');
      Udp.endPacket();
    }

    delay(10);
}
```

In the previous block of code, at first we include the `Ethernet` library. Then we declare the complete set of variables related to switch debouncing and LED handling. After these statements, we define some variables related to network features.

At first, we have to set the MAC address related to our own shield. This unique identifier is usually indicated on a sticker on your Ethernet shield. Please don't forget to put yours in the code.

Then, we set up the IP address of the Arduino. We can use any address as long as it respects the IP address schema and as long as it is reachable by our computer. That means on the same network or on another network, but with a router between both. However, be careful, as the IP address you chose has to be unique on a local network segment.

We also choose a UDP port for our communication. We are using the same definition with network parameters related to our computer, the second set of participants in the communication.

We declare a buffer to store the current received messages at each time. Notice the constant UDP_TX_PACKET_MAX_SIZE. It is defined in the Ethernet library. Basically, it is defined as 24 bytes in order to save memory. We could change that. Then, we instantiate the EthernetUDP object in order to receive and send datagram over UDP. The setup() function block contains statements for switch and LED, then for Ethernet itself.

We begin the Ethernet communication using the MAC and IP addresses. Then we open and listen at the UDP port defined in the definition, which is 9999 in our case. The loop() function seems a bit thick, but we can divide it in 2 parts.

In the first part, we check if the Arduino has received a packet. If it has, it is checked by calling the parsePacket() function of the Ethernet library and checking if that one returns a packet size different than zero. We read the data and store it into the packetBuffer variable.

Then we check if this variable equals Light or Dark and act accordingly by switching on or off the LED on the Arduino board.

In the second part, we can see the same debouncing structure as we have seen in *Chapter 5*. At the end of this part, we check if the switch is pushed or released and depending on the state send a UDP message to the computer.

Let's check the Processing/Computer part now.

Coding a Processing Applet communicating on Ethernet

Let's check the code at `Chapter11/WiredEthernetProcessing`.

We need the library hypermedia. We can find it at `http://ubaa.net/shared/processing/udp`.

```
import hypermedia.net.*;

UDP udp;  // define the UDP object
String currentMessage;

String ip       = "192.168.1.123"; // the Arduino IP address
int port        = 9999;            // the Arduino UDP port

void setup() {
  size(700, 700);
  noStroke();
  fill(0);

  udp = new UDP( this, 10000 );   // create UDP socket
  udp.listen( true );             // wait for incoming message
}

void draw()
{
  ellipse(width/2, height/2, 230, 230);
}

void receive( byte[] data ) {

  // if the message could be "Pushed" or "Released"
  if ( data.length == 6 || data.length == 8 )
  {
    for (int i=0; i < data.length; i++)
    {
      currentMessage += data[i];
    }

    // if the message is really Pushed
    // then answer back by sending "Light"
    if (currentMessage == "Pushed")
    {
      udp.send("Light", ip, port );
      fill(255);
    }
    else if (currentMessage == "Released")
    {
```

```
        udp.send("Dark", ip, port );
        fill(0);
      }
    }
  }
```

We import the library first. Then we define the UDP object and a String variable for the current received message.

Here too, we have to define the IP address of the remote participant, the Arduino. We also define the port opened and available for the communication on the Arduino side, here it is 9999.

Of course, this has to match the one defined in the Arduino firmware. In the `setup()` function, we define some drawing parameters and then instantiate the UDP socket on the UDP port 10000 and we set it to listening mode, waiting for incoming messages.

In the `draw()` function, we draw a circle. The `receive()` function is a callback used by the code when packets are incoming. We test the length of packets in bytes because we want to react to only two different messages here (`Pushed` or `Released`), so we check if the length is 6 or 8 bytes. All other packets won't be processed. We could implement a better checking mechanism, but this one works fine.

As soon as one of these lengths match, we concatenate each byte into the String variable `currentMessage`. This provides an easy way to compare the content to any other string.

Then, we compare it to `Pushed` and `Released` and act accordingly by sending back the message `Light` to the Arduino and filling our drawn circle with white color, or by sending back the message `Dark` to the Arduino and filling our drawn circle with black color.

We just designed our first basic communication protocol using Ethernet and UDP.

Some words about TCP

In my own design, I often use UDP for communication between systems. It is much lighter than TCP and is quite sufficient for our purposes.

In some cases, you would need to have the flow control provided by TCP. The Ethernet library we just used provides TCP features too. You can find the reference page at `http://arduino.cc/en/Reference/Ethernet`.

`Server` and `Client` classes can be used for this purpose especially, implementing function testing if a connection has been opened, if it is still valid, and so on.

We will learn how to connect our Arduino to some live server on the Internet at the end of this chapter.

Bluetooth communications

Bluetooth is a wireless technology standard. It provides a way to exchange data over short distances using short-wavelength radio transmissions in the band 2,400 to 2,480 MHz.

It allows to create PANs (Personal Area Networks) with the "correct" level of security. It is implemented on various types of devices such as computers, smartphones, sound systems that can read digital audio from a remote source, and so on.

Arduino BT board natively implements this technology. It is now supplied with ATmega328 and a Bluegiga WT11 Bluetooth module. The reference page is `http://www.arduino.cc/en/Main/ArduinoBoardBluetooth`.

In my opinion, the best way to proceed in many projects is to keep a general purpose board at the core of our designs and to add new features by adding only what we need as external modules. Following this, we are going to use the Arduino UNO R3 here with an external Bluetooth module.

We are going to make a small project using Processing again. You click somewhere over the Processing canvas and the Processing applet sends a message over Bluetooth to the Arduino, which reacts by switching its built-in LED on or off.

Wiring the Bluetooth module

Check the following figures:

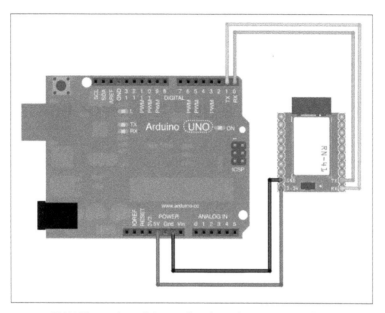

RN41 Bluetooth module wired to the Arduino via a serial link

The corresponding circuit diagram is as follows:

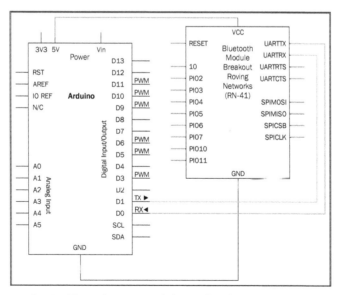

Roving Networks RN41 module wired to the Arduino board

There is a Roving Networks RN41 Bluetooth module wired to the Arduino board.

You can find it at `https://www.sparkfun.com/products/10559`.

Here we are using the basic serial link communication between the Arduino itself and the Bluetooth module.

We suppose our computer has Bluetooth capabilities and that those are activated.

Coding the firmware and the Processing applet

The firmware is as follows. You can find it in at `Chapter11/Bluetooth`.

```
// LED stuff
const int ledPin =  13;        // pin of the board built-in LED

void setup() {
  pinMode(ledPin, OUTPUT);    // the led pin is setup as an output

  Serial.begin(9600);          // start serial communication at 9600bps
}

void loop()
{
  if (Serial.available() > 0) {

    incomingByte = Serial.read();

    if (incomingByte == 1) digitalWrite(ledPin, HIGH);
    else if (incomingByte == 0) digitalWrite(ledPin, LOW);
  }
}
```

We basically instantiate the `Serial` communication with the Bluetooth module, then we check if any bytes are available from it and parse them. If a message is available and equals 1, we switch on the LED; if it equals 0, we switch off the LED.

The processing code is as follows:

```
import processing.serial.*;
Serial port;

int bgcolor, fgcolor;

void setup() {
```

```
    size(700, 700);
    background(0);
    stroke(255);
    bgcolor = 0;
    fgcolor = 255;

    println(Serial.list());
    port = new Serial(this, Serial.list()[2], 9600);

}
void draw() {
  background(bgcolor);
  stroke(fgcolor);
  fill(fgcolor);
  rect(100, 100, 500, 500);
}

void mousePressed() {
  if (mouseX > 100 && mouseX < 600 && mouseY > 100 && mouseY < 600)
  {
      bgcolor = 255;
      fgcolor = 0;
      port.write('1');
  }
}

void mouseReleased() {

      bgcolor = 0;
      fgcolor = 255;
      port.write('0');
}
```

We first include the serial library. In the `setup()` function, we define some drawing bits, then we print the list of serial device to the Processing log area. This displays a list and we have to find the right Bluetooth module of our computer. In my case, this was the third one and I used this to instantiate the `Serial` communication in the latest statement of the `setup()` function:

```
port = new Serial(this, Serial.list()[2], 9600);
```

The `draw()` function only sets up:

- Background color according to the variable `bgcolor`

- Stroke color according to the variable `fgcolor`

- Fill color according to the variable `fgcolor`

Then we draw a square.

The `mousePressed()` and `mouseReleased()` functions are two Processing callbacks respectively which are called when a mouse event occurs, when you push a button on the mouse and release it.

As soon as the mouse is pressed, we check where the cursor was when it was pressed. In my case, I defined the area inside the square.

If we press the button in the square, a visual feedback occurs in order to tell us the order has been received, but the most important thing is the `digitalWrite('1')` function of course.

We write the value 1 to the Bluetooth module.

In the same way, as soon as we release the mouse button, a "0" is written to the Bluetooth module of the computer. Of course, these messages are sent to the Arduino and the latter switches the LED on or off.

We just checked a nice example of an external module providing wireless Bluetooth communication feature to the Arduino.

As we noticed, we don't have to use a particular library for this purpose because the module itself is able to connect and send/receive data by itself only if we send serial data to it. Indeed, the communication between Arduino and the module is a basic serial one.

Let's improve our data communication over the air using Ethernet Wi-Fi.

Playing with Wi-Fi

We previously learned how to use the Ethernet library. Then, we tested Bluetooth for short-range network communications. Now, let's test Wi-Fi for medium range communications still without any wire.

What is Wi-Fi?

Wi-Fi is a set of communication protocols wireless driven by standards of IEEE 802.11. These standards describe characteristics of Wireless Local Area Networks (WLANs).

Basically, multiple hosts having Wi-Fi modules can communicate using their IP stacks without wire. There are multiple networking modes used by Wi-Fi.

Infrastructure mode

In that mode, Wi-Fi hosts can communicate between each other via an access point.

This access point and hosts have to be set up with the same **Service Set Identifier** (SSID), which is a network name used as a reference.

This mode is interesting because it provides security by the fact that each host has to pass by the access point in order to access the global network. We can configure some access lists in order to control which host can connect and which cannot.

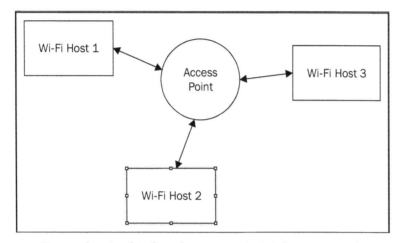

Hosts exchanging data through an access point in infrastructure mode

Ad hoc mode

In this mode, each host can connect to each one directly without access points. It is very useful to quickly connect two hosts in order to share documents and exchange data.

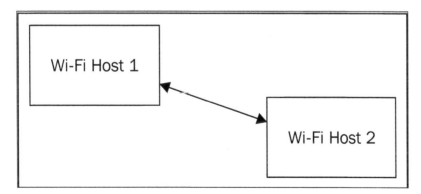

Two hosts directly connected in ad hoc mode

Other modes

There are also two other modes. **Bridge mode** is a way to link multiple access points. We can imagine a work group sparse in two buildings; we could use two different access points and connect them together using a bridge mode.

There is also a trivial mode named **range-extender mode**. It is used to repeat the signal and provide a connection between two hosts, two access points or a host, and an access point when those are too far.

The Arduino Wi-Fi shield

This shield adds the wireless networking capabilities to the Arduino board. The official shield also contains an SD card slot providing storing features too. It provides:

- Connection via 802.11b/g networks
- Encryption using WEP or WPA2 personal
- FTDI connection for serial debugging of the shield itself
- Mini-USB to update the Wi-Fi shield firmware itself

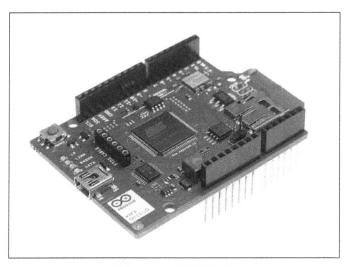

The Arduino Wi-Fi Shield

It is based on the HDG104 Wireless LAN 802.11b/g system in-package. A proper Atmega 32 UC3 provides the network IP stack.

A dedicated native library named **WiFi library** provides all that we need to connect our board wireless to any network. The reference is provided at `http://arduino.cc/en/Reference/WiFi`.

This shield is available from many distributors as well as from the Arduino store: `http://store.arduino.cc/ww/index.php?main_page=product_info&cPath=11_5&products_id=237`.

Let's try to connect our Arduino to our Wi-Fi network.

Basic Wi-Fi connection without encryption

Here, we don't have to draw any schematic. Basically, we connect the shield to the Arduino and upload our code to it. We are going to test a basic connection without encrypting anything, at first.

The Accept Point has to provide a DHCP server; the latter will deliver an IP address to our Arduino-based system.

Let's check the example `ConnectNoEncryption` provided with the `WiFi` library.

```
#include <WiFi.h>

char ssid[] = "yourNetwork";      // the name of your network
int status = WL_IDLE_STATUS;      // the Wifi radio's status

void setup() {
  //Initialize serial and wait for port to open:
  Serial.begin(9600);

  // check for the presence of the shield:
  if (WiFi.status() == WL_NO_SHIELD) {
    Serial.println("WiFi shield not present");
    // don't continue:
    while(true)
delay(30) ;
  }

  // attempt to connect to Wifi network:
  while ( status != WL_CONNECTED) {
    Serial.print("Attempting to connect to open SSID: ");
    Serial.println(ssid);
    status = WiFi.begin(ssid);

    // wait 10 seconds for connection:
```

```
    delay(10000);
  }

  // you're connected now, so print out the data:
  Serial.print("You're connected to the network");
  printCurrentNet();
  printWifiData();
}

void loop() {
  // check the network connection once every 10 seconds:
  delay(10000);
  printCurrentNet();
}

void printWifiData() {
  // print your WiFi shield's IP address:
  IPAddress ip = WiFi.localIP();
  Serial.print("IP Address: ");
  Serial.println(ip);
  Serial.println(ip);

  // print your MAC address:
  byte mac[6];
  WiFi.macAddress(mac);
  Serial.print("MAC address: ");
  Serial.print(mac[5],HEX);
  Serial.print(":");
  Serial.print(mac[4],HEX);
  Serial.print(":");
  Serial.print(mac[3],HEX);
  Serial.print(":");
  Serial.print(mac[2],HEX);
  Serial.print(":");
  Serial.print(mac[1],HEX);
  Serial.print(":");
  Serial.println(mac[0],HEX);

  // print your subnet mask:
  IPAddress subnet = WiFi.subnetMask();
  Serial.print("NetMask: ");
  Serial.println(subnet);

  // print your gateway address:
```

```
    IPAddress gateway = WiFi.gatewayIP();
    Serial.print("Gateway: ");
    Serial.println(gateway);
}

void printCurrentNet() {
  // print the SSID of the network you're attached to:
  Serial.print("SSID: ");
  Serial.println(WiFi.SSID());

  // print the MAC address of the router you're attached to:
  byte bssid[6];
  WiFi.BSSID(bssid);
  Serial.print("BSSID: ");
  Serial.print(bssid[5],HEX);
  Serial.print(":");
  Serial.print(bssid[4],HEX);
  Serial.print(":");
  Serial.print(bssid[3],HEX);
  Serial.print(":");
  Serial.print(bssid[2],HEX);
  Serial.print(":");
  Serial.print(bssid[1],HEX);
  Serial.print(":");
  Serial.println(bssid[0],HEX);

  // print the received signal strength:
  long rssi = WiFi.RSSI();
  Serial.print("signal strength (RSSI):");
  Serial.println(rssi);

  // print the encryption type:
  byte encryption = WiFi.encryptionType();
  Serial.print("Encryption Type:");
  Serial.println(encryption,HEX);
}
```

At first, we include the `WiFi` library. Then, we set the name of our network, the SSID. Please be careful to change it to your own SSID.

In the `setup()` function, we instantiate the `Serial` connection. Then, we check the presence of the shield by calling the function `WiFi.status()`.

If the latter returns the value `WL_NO_SHIELD` (which is a constant defined inside the WiFi library), that means there is no shield. In that case, an infinite loop is executed with a `while(true)` statement without the `break` keyword inside.

If it returns a value different than WL_CONNECTED, then we print a statement in order to inform that it is trying to connect. Then, WiFi.begin() tries to connect. This is a usual structure providing a way to try to connect while it isn't connected, constantly, and each 10 s considering the delay() function is called.

Then, if the connection occurs, the status becomes WL_CONNECTED, we exit from the while loop and continue.

There is something printed to serial too, saying the board has achieved connection status.

We also call two functions. These functions print to serial many elements related to network parameters and status. I'll let you discover each of them using the http://arduino.cc/en/Reference/WiFi reference quoted earlier.

After this connection, we can begin to exchange data. As you probably know, using Wi-Fi especially without security can lead to problems. Indeed, it is very easy to capture packets from an unprotected Wi-Fi network.

Let's use the WiFi library with more security.

Arduino Wi-Fi connection using WEP or WPA2

If you open both code ConnectWithWEP and ConnectWithWPA, there are minor differences with the preceding example.

Using WEP with Wi-Fi library

If we use a 40-bit WEP, we need a key containing 10 characters that must be hexadecimal. If we use 128-bit WEP, we need a key containing 26 characters, also hexadecimal. This key must be specified within the code.

We replaced the call to WiFi.begin(), which had only one argument, by two new arguments related to WEP encryption. This is the only difference.

For many reasons that we won't discuss here, WEP is considered too weak in terms of security, so most people and organizations have moved to the more secure WPA2 alternative.

Using WPA2 with Wi-Fi library

Following the same schema, we need only a password here. Then, we call WiFi.begin() with 2 arguments: the SSID and the password.

In both cases we just checked, we only had to pass some additional arguments with WiFi.begin() in order to secure things a bit more.

Arduino has a (light) web server

Here, we use the code `WifiWebServer` provided with the library.

In this example, Arduino acts as a web server after having been connected to a WEP or WPA Wi-Fi network.

```
#include <WiFi.h>

char ssid[] = "yourNetwork";      //  your network SSID (name)
char pass[] = "secretPassword";   // your network password
int keyIndex = 0;                 // your network key Index number
(needed only for WEP)

int status = WL_IDLE_STATUS;

WiFiServer server(80);

void setup() {
  //Initialize serial and wait for port to open:
  Serial.begin(9600);
  while (!Serial) {
    ; // wait for serial port to connect. Needed for Leonardo only
  }

  // check for the presence of the shield:
  if (WiFi.status() == WL_NO_SHIELD) {
    Serial.println("WiFi shield not present");
    // don't continue:
    while(true);
  }

  // attempt to connect to Wifi network:
  while ( status != WL_CONNECTED) {
    Serial.print("Attempting to connect to SSID: ");
    Serial.println(ssid);
    // Connect to WPA/WPA2 network. Change this line if using open or
WEP network:
    status = WiFi.begin(ssid, pass);

    // wait 10 seconds for connection:
    delay(10000);
  }
  server.begin();
  // you're connected now, so print out the status:
  printWifiStatus();
}
```

```
void loop() {
  // listen for incoming clients
  WiFiClient client = server.available();
  if (client) {
    Serial.println("new client");
    // an http request ends with a blank line
    boolean currentLineIsBlank = true;
    while (client.connected()) {
      if (client.available()) {
        char c = client.read();
        Serial.write(c);
        // if you've gotten to the end of the line (received a newline
        // character) and the line is blank, the http request has
ended,
        // so you can send a reply
        if (c == '\n' && currentLineIsBlank) {
          // send a standard http response header
          client.println("HTTP/1.1 200 OK");
          client.println("Content-Type: text/html");
          client.println("Connnection: close");
          client.println();
          client.println("<!DOCTYPE HTML>");
          client.println("<html>");
          // add a meta refresh tag, so the browser pulls again every
5 seconds:
          client.println("<meta http-equiv=\"refresh\"
content=\"5\">");
          // output the value of each analog input pin
          for (int analogChannel = 0; analogChannel < 6;
analogChannel++) {
            int sensorReading = analogRead(analogChannel);
            client.print("analog input ");
            client.print(analogChannel);
            client.print(" is ");
            client.print(sensorReading);
            client.println("<br />");
          }
          client.println("</html>");
           break;
        }
        if (c == '\n') {
          // you're starting a new line
          currentLineIsBlank = true;
        }
        else if (c != '\r') {
```

```
                    // you've gotten a character on the current line
                    currentLineIsBlank = false;
                }
            }
        }
        // give the web browser time to receive the data
        delay(1);
            // close the connection:
            client.stop();
            Serial.println("client disonnected");
    }
}


void printWifiStatus() {
    // print the SSID of the network you're attached to:
    Serial.print("SSID: ");
    Serial.println(WiFi.SSID());

    // print your WiFi shield's IP address:
    IPAddress ip = WiFi.localIP();
    Serial.print("IP Address: ");
    Serial.println(ip);

    // print the received signal strength:
    long rssi = WiFi.RSSI();
    Serial.print("signal strength (RSSI):");
    Serial.print(rssi);
    Serial.println(" dBm");
}
```

Let's explain the underlying concepts in these statements.

We explain only the new part of the code, not the autoconnect and encryption statements, because we did that earlier.

The `WiFiServer server(80)` statement instantiates a server on a specific port. Here, the TCP port chosen is 80, the standard HTTP server TCP port.

In the `setup()` function, we auto-connect the Arduino to the Wi-Fi network, then we start the server. Basically, it opens a socket on TCP port 80 and begins to listen on this port.

In the `loop()` function, we check if there is an incoming client to our web server embedded on the Arduino. This is done with `WiFiClient client = server. available();`

Then, we have a condition on client instance. If there is no client, we basically do nothing, and execute the loop again until we have a client.

As soon as we have one, we print this to serial in order to give feedback. Then, we check if the client is effectively connected and if there is data in the reading buffer. We then print this data if it is available and answer the client by sending the standard HTTP response header. This is done basically by printing bytes to the client instance itself.

The code includes some dynamic features and sends some values read on the board itself like the value coming from the ADC of each analog input.

We could try to connect some sensors and provide values of each of them through a webpage directly handled by the Arduino itself. I'll let you check the other part of the code. This deals with standard HTTP messages.

Tweeting by pushing a switch

Connecting the Arduino to networks obviously brings the Internet to mind. We could try to create a small system that can send messages over the Internet. I choose to use the micro-blogging service Twitter because it provides a nice communication API.

We are going to use the same circuit that we used in the *Wiring Arduino to wired Ethernet* section except that here we are using the Arduino MEGA related to some memory constraints with a smaller board.

An overview of APIs

API stands for **Application Programming Interface**. Basically, it defines ways to exchange data with the considered system. We can define APIs in our systems in order to make them communicate with others.

For instance, we could define an API in our Arduino firmware that would explain how and what to send in order to make the LED on the board switch on and off. We won't describe the whole firmware, but we would provide to the world a basic document explaining precisely the format and data to send from the Internet, for instance, to use it remotely. That would be an API.

Twitters API

Twitter, as do many other social network-related systems on the Internet, provides an API. Other programmers can use it to get data and send data too. All data specifications related to Twitters API are available at `https://dev.twitter.com`.

In order to use the API, we have to create an application on Twitters developer website. There are some special security parameters to set up, and we have to agree upon some rules of use that respect data requests rate and other technical specifications.

We can create an application by going to `https://dev.twitter.com/apps/new`.

That will provide us with some credential information, in particular an access token and a token secret. These are strings of characters that have to be used following some protocols to be able to access the API.

Using the Twitter library with OAuth support

Markku Rossi created a very powerful and reliable library embedding the OAuth support and intended for sending tweets directly from the Arduino. The official library website is `http://www.markkurossi.com/ArduinoTwitter`.

This library needs to be used with a board with more than the usual amount of memory. The Arduino MEGA runs it perfectly.

OAuth is an open protocol to allow secure authorization in a simple and standard method from web, mobile, and desktop applications. This is defined at `http://oauth.net`.

Basically, this is a way to enable third-party application to obtain limited access to an HTTP service. By sending some specific string of characters, we can grant access to a host and make it communicate with the API.

This is what we are going to do together as a nice example that you could reuse for other APIs on the Web.

Grabbing credentials from Twitter

Markku's library implements the OAuth request signing, but it doesn't implement the OAuth Access Token retrieval flow. We can retrieve our token by using this guide on the Twitter website where we created our application: `https://dev.twitter.com/docs/auth/tokens-devtwittercom`.

You need to keep handy the Access token and Access token secret, as we are going to include them in our firmware.

Coding a firmware connecting to Twitter

Markku's library is easy to use. Here is a possible code connecting the Arduino to your Ethernet network so that you can tweet messages directly.

You can find it at `Chapter11/tweetingButton/`.

```
#include <SPI.h>
#include <Ethernet.h>
#include <sha1.h>
#include <Time.h>
#include <EEPROM.h>
#include <Twitter.h>

// Switch
const int switchPin = 2;
int switchState = 0;
int lastSwitchState = LOW;
long lastDebounceTime = 0;
long debounceDelay = 50;


// Local network configuration
uint8_t mac[6] =       {
  0xc4, 0x2c, 0x03, 0x0a, 0x3b, 0xb5};     // USE YOUR MAC ADDRESS
IPAddress ip(192, 168, 1, 43);             // USE IP ON YOUR NETWORK
IPAddress gateway(192, 168, 1, 1);         // USE YOUR GATWEWAY IP
ADDRESS
IPAddress subnet(255, 255, 255, 0);        // USE YOUR SUBNET MASK

// IP address to Twitter
IPAddress twitter_ip(199, 59, 149, 232);
uint16_t twitter_port = 80;

unsigned long last_tweet = 0;
#define TWEET_DELTA (60L * 60L)

// Store the credentials
const static char consumer_key[] PROGMEM = "xxxxxxxxxxxxx";
const static char consumer_secret[] PROGMEM
= "yyyyyyyyyyyyy";

#DEFINE ALREADY_TOKENS 0 ; // Change it at 1 when you put your tokens

char buffer[512];
Twitter twitter(buffer, sizeof(buffer));


void setup() {
  Serial.begin(9600);
  Serial.println("Arduino Twitter demo");

  // the switch pin is setup as an input
  pinMode(switchPin, INPUT);
```

```
  // start the network connection
  Ethernet.begin(mac, ip, dns, gateway, subnet);

  // define twitter entry point
  twitter.set_twitter_endpoint(PSTR("api.twitter.com"),
  PSTR("/1/statuses/update.json"),
  twitter_ip, twitter_port, false);
  twitter.set_client_id(consumer_key, consumer_secret);

  // Store or read credentials in EEPROM part of the board
#if ALREADY_TOKENS
  /* Read OAuth account identification from EEPROM. */
  twitter.set_account_id(256, 384);
#else
  /* Set OAuth account identification from program memory. */
  twitter.set_account_id(PSTR("*** set account access token here
***"),
  PSTR("*** set account token secret here ***"));
#endif

  delay(500);
}

void loop() {
  if (twitter.is_ready()) // if the twitter connection is okay
  {
    unsigned long now = twitter.get_time();
    if (last_tweet == 0) last_tweet = now - TWEET_DELTA + 15L;

    // read the state of the digital pin
    int readInput = digitalRead(switchPin);
    if (readInput != lastSwitchState)
    {
      lastDebounceTime = millis();
    }

    if ( (millis() - lastDebounceTime) > debounceDelay )
    {
      switchState = readInput;
    }

    lastSwitchState = readInput;
    if (switchState == HIGH)  // if you push the button
    {
      if (now > last_tweet + TWEET_DELTA) // if you didn't tweet for a
while
      {

          char msg[32];
```

```
        sprintf(msg, "Tweeting from #arduino by pushing a button is
cool, thanks to @julienbayle");

        // feedback to serial monitor
        Serial.print("Posting to Twitter: ");
        Serial.println(msg);

        last_tweet = now;

        if (twitter.post_status(msg))
          Serial.println("Status updated");
        else
          Serial.println("Update failed");
      }
      else Serial.println("Wait a bit before pushing it again!");
    }
  }
  delay(5000); // waiting a bit, just in case
}
```

Let's explain things here. Please note, this is a code including many things we already discovered and learned together:

- Button push with debouncing system
- Ethernet connection with the Arduino Ethernet Shield
- Twitter library example

We first include a lot of library headers:

- SPI and Ethernet for network connection
- Sha1 for credentials encryption
- Time for time and date specific functions used by Twitter library
- EEPROM to store credentials in EEPROM of the board
- Twitter library itself

Then, we include the variable related to the button itself and the debouncing system.

We configure the network parameters. Please notice you have to put your own elements here, considering your network and Ethernet shield. Then, we define the IP address of Twitter.

We define the TWEET_DELTA constant for further use, with respect to the Twitter API use that forbids us from sending too many tweets at a time. Then, we store our credentials. Please use yours, related to the application you created on the Twitter website for our purpose. At last we create the object twitter.

In the setup() function, we start the Serial connection in order to send some feedback to us. We configure the digital pin of the switch and start the Ethernet connection. Then, we have all the wizardry about Twitter. We first choose the entry point defined by the Twitter API docs itself. We have to put our Access token and Token secret here too. Then, we have a compilation condition: #if TOKEN_IN_MEMORY.

TOKEN_IN_MEMORY is defined before as 0 or 1. Depending on its value, the compilation occurs in one manner or another.

In order to store credentials to the EEPROM of the board, we first have to put the value 0. We compile it and run it on the board. The firmware runs and writes the tokens in memory. Then, we change the value to 1 (because tokens are now in memory) and we compile it and run it on the board. From now, the firmware will read credentials from EEPROM.

Then, the loop() function is quite simple considering what we learned before.

We first test if the twitter connection to the API is okay. If it is okay, we store the time and the time of the last tweet at an initial value. We read the debounce value of the digital input.

If we push the button, we test to see if we did that in less than the TWEET_DELTA amount of time. If it is the case, we are safe with respect to the Twitter API rules and we can tweet.

At last, we store a message in the char array msg. And we tweet the message by using twitter.post_status() function. While using it, we also test what it returns. If it returns 1, it means the tweet occurred. That provides this information to the user through serial monitor.

All API providers work in the same way. Here, we were very helped by the Twitter library we used, but there are other libraries also for other services on the Internet. Also, each service provides the complete documentation to use their API. Facebook API resources are available here: https://developers.facebook.com/. Google+ API resources are available here: https://developers.google.com/+/api/. Instagram API resources are available here: http://instagram.com/developer. And we could find a lot of others.

Summary

In this chapter, we learned how to extend the area of communication of our Arduino boards. We were used to making very local connections; we are now able to connect our board to the Internet and potentially communicate with the whole planet.

We described Wired Ethernet, Wi-Fi, Bluetooth connections, and how to use Twitters API.

We could have described the Xbee board, which uses radio frequencies, too, but I preferred to describe IP-related stuff because I consider them to be the safest way to transmit data. Of course, Xbees shield solution is a very nice one too and I used it myself in many projects.

In the next chapter, we are going to describe and dig into the Max 6 framework. This is a very powerful programming tool that can generate and parse data and we are going to explain how we can use it with Arduino.

12

Playing with
Max 6 Framework

This chapter will teach us some tips and techniques that we can use with the Max 6 graphical programming framework and Arduino boards.

We introduced this amazing framework in *Chapter 6, Sense the World – Feeling with Analog Inputs*, while we learned about Arduino analog input handling. Reading the previous chapter is a requirement to better understand and learn the techniques developed in this chapter. I even suggest you read the Max 6 introduction part again.

In this chapter, we will learn how to send data to Arduino from Max 6. We will also describe how we can handle and parse the data being received from Arduino.

Arduino adds a lot of features to your Max 6 programs. Indeed, it provides a way to plug Max 6 into the real physical world. Through two examples, we are going to understand a nice way of working with Arduino, the computer and most advanced programming framework ever.

Let's go.

Communicating easily with Max 6 – the [serial] object

As we already discussed in *Chapter 6, Sensing the World – Feeling with Analog Inputs*, the easiest way to exchange data between your computer running a Max 6 patch and your Arduino board is via the serial port. The USB connector of our Arduino boards includes the FTDI integrated circuit EEPROM FT-232 that converts the RS-232 plain old serial standard to USB.

We are going to use again our basic USB connection between Arduino and our computer in order to exchange data here.

The [serial] object

We have to remember the [serial] object's features. It provides a way to send and receive data from a serial port. To do this, there is a basic patch including basic blocks. We are going to improve it progressively all along this subchapter.

The [serial] object is like a buffer we have to poll as much as we need. If messages are sent from Arduino to the serial port of the computer, we have to ask the [serial] object to pop them out. We are going to do this in the following pages.

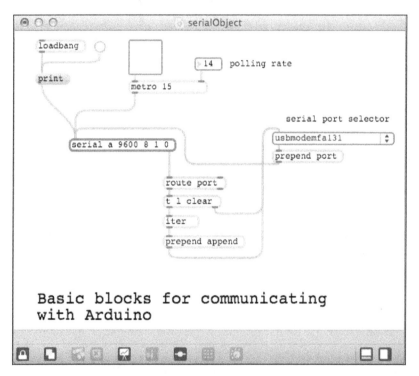

You can find it in the Chapter12 folder; the patch is named serialObject.maxpat.

Of course, this chapter is also a pretext for me to give you some of my tips and tricks in Max 6 itself. Take them and use them; they will make your patching life easier.

Selecting the right serial port

In *Chapter 6, Sense the World – Feeling with Analog Inputs,* we used the message
(print) sent to [serial] in order to list all the serial ports available on the
computer. Then we checked the Max window. That was not the smartest solution.
Here, we are going to design a better one.

We have to remember the [loadbang] object. It fires a bang, that is, a (print)
message to the following object as soon as the patch is loaded. It is useful to set
things up and initialize some values as we could inside our setup() block in our
Arduino board's firmware.

Here, we do that in order to fill the serial port selector menu. When the [serial]
object receives the (print) message, it pops out a list of all the serial ports available on
the computer from its right outlet prepended by the word port. We then process the
result by using [route port] that only parses lists prepended with the word port.

The [t] object is an abbreviation of [trigger]. This object sends the incoming
message to many locations, as is written in the documentation, if you assume the use
of the following arguments:

- b means bang
- f means float number
- i means integer
- s means symbol
- l means list (that is, at least one element)

We can also use constants as arguments and as soon as the input is received, the
constant will be sent as it is.

At last, the [trigger] output messages in a particular order: from the rightmost
outlet to the leftmost one.

So here we take the list of serial ports being received from the [route] object; we
send the clear message to the [umenu] object (the list menu on the left side) in order
to clear the whole list. Then the list of serial ports is sent as a list (because of the first
argument) to [iter]. [iter] splits a list into its individual elements.

[prepend] adds a message in front of the incoming input message.

That means the global process sends messages to the [umenu] object similar to the following:

- append xxxxxx
- append yyyyyy

Here xxxxxx and yyyyyy are the serial ports that are available.

This creates the serial port selector menu by filling the list with the names of the serial ports. This is one of the typical ways to create some helpers, in this case the menu, in our patches using UI elements.

As soon as you load this patch, the menu is filled, and you only have to choose the right serial port you want to use. As soon as you select one element in the menu, the number of the element in the list is fired to its leftmost outlet. We prepend this number by port and send that to [serial], setting it up to the right-hand serial port.

Polling system

One of the most used objects in Max 6 to send regular bangs in order to trigger things or count time is [metro].

We have to use one argument at least; this is the time between two bangs in milliseconds.

Banging the [serial] object makes it pop out the values contained in its buffer.

If we want to send data continuously from Arduino and process them with Max 6, activating the [metro] object is required. We then send a regular bang and can have an update of all the inputs read by Arduino inside our Max 6 patch.

Choosing a value between 15 ms and 150 ms is good but depends on your own needs.

Let's now see how we can read, parse, and select useful data being received from Arduino.

Parsing and selecting data coming from Arduino

First, I want to introduce you to a helper firmware inspired by the *Arduino2Max* page on the Arduino website but updated and optimized a bit by me. It provides a way to read all the inputs on your Arduino, to pack all the data read, and to send them to our Max 6 patch through the [serial] object.

The readAll firmware

The following code is the firmware. You can find it in `Chapter12/ReadAll`:

```
int val = 0;

void setup()
{
  Serial.begin(9600);
  pinMode(13,INPUT);
}

void loop()
{
  // Check serial buffer for characters incoming
  if (Serial.available() > 0){

    // If an 'r' is received then read all the pins
    if (Serial.read() == 'r') {

      // Read and send analog pins 0-5 values
      for (int pin= 0; pin<=5; pin++){
        val = analogRead(pin);
        sendValue (val);
      }

      // Read and send digital pins 2-13 values
      for (int pin= 2; pin<=13; pin++){
        val = digitalRead(pin);
        sendValue (val);
      }

      Serial.println();// Carriage return to mark end of data flow.
      delay (5);      // prevent buffer overload

    }

  }
}

void sendValue (int val){
  Serial.print(val);
  Serial.write(32);   // add a space character after each value sent
}
```

For starters, we begin the serial communication at 9600 bauds in the `setup()` block.

As usual with serial communication handling, we check if there is something in the serial buffer of Arduino at first by using the `Serial.available()` function. If something is available, we check if it is the character r. Of course, we can use any other character. r here stands for read, which is basic. If an r is received, it triggers the read of both analog and digital ports. Each value (the `val` variable) is passed to the `sendValue()` function; this basically prints the value into the serial port and adds a space character in order to format things a bit to provide an easier parsing by Max 6. We could easily adapt this code to only read some inputs and not all. We could also remove the `sendValue()` function and find another way of packing data.

At the end, we push a carriage return to the serial port by using `Serial.println()`. This creates a separator between each pack of data that is sent.

Now, let's improve our Max 6 patch to handle this pack of data being received from Arduino.

The ReadAll Max 6 patch

The following screenshot is the `ReadAll` Max patch that provides a way to communicate with our Arduino:

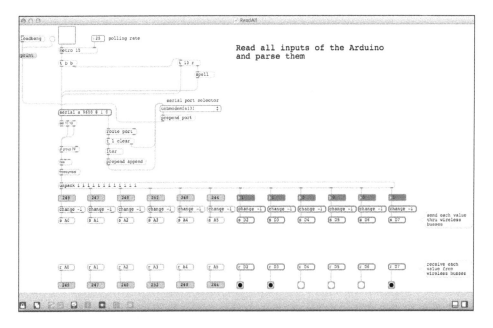

You can find this patch in the `Chapter12` folder. It is named `ReadAll.maxpat`.

We added a couple of more things to the basic building blocks in the previous patch.

Requesting data from Arduino

First, we will see a [t b b] object. It is also a trigger, ordering bangs provided by the [metro] object. Each bang received triggers another bang to another [trigger] object, then another one to the [serial] object itself.

The [t 13 r] object can seem tricky. It just triggers a character r and then the integer 13. The character r is sent to [spell] that converts it to ASCII code and then sends the result to [serial]. 13 is the ASCII code for a carriage return.

This structure provides a way to fire the character r to the [serial] object, which means to Arduino, each time that the metro bangs. As we already see in the firmware, it triggers Arduino to read all its inputs, then to pack the data, and then to send the pack to the serial port for the Max 6 patch.

To summarize what the metro triggers at each bang, we can write this sequence:

1. Send the character r to Arduino.
2. Send a carriage return to Arduino.
3. Bang the [serial] object.

This triggers Arduino to send back all its data to the Max patch.

Parsing the received data

Under the [serial] object, we can see a new structure beginning with the [sel 10 13] object. This is an abbreviation for the [select] object. This object selects an incoming message and fires a bang to the specific output if the message equals the argument corresponding to the specific place of that output. Basically, here we select 10 or 13. The last output pops the incoming message out if that one doesn't equal any argument.

Here, we don't want to consider a new line feed (ASCII code 10). This is why we put it as an argument, but we don't do anything if that's the one that has been selected. It is a nice trick to avoid having this message trigger anything and even to not have it from the right output of [select].

Here, we send all the messages received from Arduino, except 10 or 13, to the [zl group 78] object. The latter is a powerful list for processing many features. The group argument makes it easy to group the messages received in a list. The last argument is to make sure we don't have too many elements in the list. As soon as [zl group] is triggered by a bang *or* the list length reaches the length argument value, it pops out the whole list from its left outlet.

Here, we "accumulate" all the messages received from Arduino, and as soon as a carriage return is sent (remember we are doing that in the last rows of the loop() block in the firmware), a bang is sent and all the data is passed to the next object.

We currently have a big list with all the data inside it, with each value being separated from the other by a space character (the famous ASCII code 32 we added in the last function of the firmware).

This list is passed to the [itoa] object. **itoa** stands for *integer to ASCII*. This object converts integers to ASCII characters.

The [fromsymbol] object converts a symbol to a list of messages.

Finally, after this [fromsymbol] object we have our big list of values separated by spaces and totally readable.

We then have to unpack the list. [unpack] is a very useful object that provides a way to cut a list of messages into individual messages. We can notice here that we implemented exactly the opposite process in the Arduino firmware while we packed each value into a big message.

[unpack] takes as many arguments as we want. It requires knowing about the exact number of elements in the list sent to it. Here we send 12 values from Arduino, so we put 12 i arguments. i stands for *integer*. If we send a float, [unpack] would cast it as an integer. It is important to know this. Too many students are stuck with troubleshooting this in particular.

We are only playing with the integer here. Indeed, the ADC of Arduino provides data from 0 to 1023 and the digital input provides 0 or 1 only.

We attached a number box to each output of the [unpack] object in order to display each value.

Then we used a [change] object. This latter is a nice object. When it receives a value, it passes it to its output only if it is different from the previous value received. It provides an effective way to avoid sending the same value each time when it isn't required.

Here, I chose the argument -1 because this is not a value sent by the Arduino firmware, and I'm sure that the first element sent will be parsed.

So we now have all our values available. We can use them for different jobs.

But I propose to use a smarter way, and this will also introduce a new concept.

Distributing received data and other tricks

Let's introduce here some other tricks to improve our patching style.

Cordless trick

We often have to use some data in our patches. The same data has to feed more than one object.

A good way to avoid messy patches with a lot of cord and wires everywhere is to use the [send] and [receive] objects. These objects can be abbreviated with [s] and [r], and they generate communication buses and provide a wireless way to communicate inside our patches.

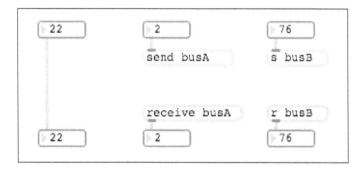

These three structures are equivalent.

The first one is a basic cord. As soon as we send data from the upper number box, it is transmitted to the one at the other side of the cord.

The second one generates a data bus named busA. As soon as you send data into [send busA], each [receive busA] object in your patch will pop out that data.

The third example is the same as the second one, but it generates another bus named busB.

This is a good way to distribute data.

I often use this for my master clock, for instance. I have one and only one master clock banging a clock to [send masterClock], and wherever I need to have that clock, I use [receive masterClock] and it provides me with the data I need.

If you check the global patch, you can see that we distribute data to the structures at the bottom of the patch. But these structures could also be located elsewhere. Indeed, one of the strengths of any visual programming framework such as Max 6 is the fact that you can visually organize every part of your code exactly as you want in your patcher. And please, do that as much as you can. This will help you to support and maintain your patch all through your long development months.

Check the previous screenshot. I could have linked the [r A1] object at the top left corner to the [p process03] object directly. But maybe this will be more readable if I keep the process chains separate. I often work this way with Max 6.

This is one of the multiple tricks I teach in my Max 6 course. And of course, I introduced the [p] object, that is the [patcher] abbreviation.

Let's check a couple of tips before we continue with some good examples involving Max 6 and Arduino.

Encapsulation and subpatching

When you open Max 6 and go to **File | New Patcher**, it opens a blank patcher. The latter, if you recall, is the place where you put all the objects. There is another good feature named **subpatching**. With this feature, you can create new patchers inside patchers, and embed patchers inside patchers as well.

A patcher contained inside another one is also named a subpatcher.

Let's see how it works with the patch named ReadAllCutest.maxpat.

[HII

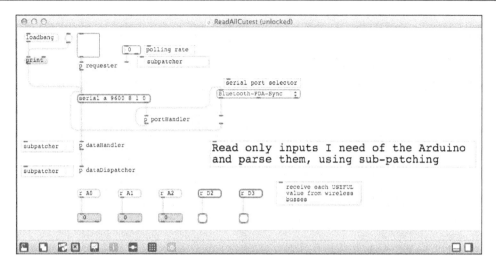

There are four new objects replacing the whole structures we designed before.

These objects are subpatchers. If you double-click on them in **patch lock mode** or if you push the *command* key (or *Ctrl* for Windows), double-click on them in **patch edit mode** and you'll open them. Let's see what is there inside them.

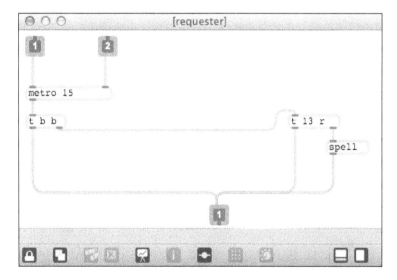

The [requester] subpatcher contains the same architecture that we designed before, but you can see the brown **1** and **2** objects and another blue **1** object. These are inlets and outlets. Indeed, they are required if you want your subpatcher to be able to communicate with the patcher that contains it. Of course, we could use the [send] and [receive] objects for this purpose too. We are going to see that in the following pages.

The position of these inlets and outlets in your subpatcher matters. Indeed, if you move the **1** object to the right of the **2** object, the numbers get swapped! And the different inlets in the upper patch get swapped too. You have to be careful about that. But again, you can organize them exactly as you want and need.

Check the next screenshot:

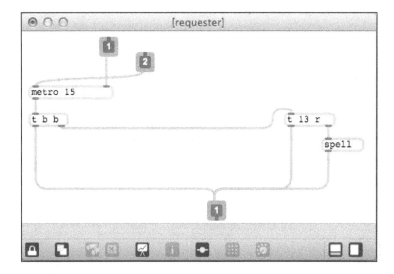

And now, check the root patcher containing this subpatcher. It automatically inverts the inlets, keeping things relevant.

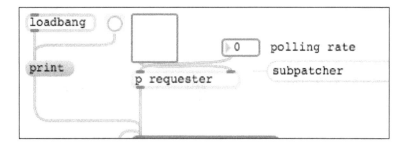

Let's now have a look at the other subpatchers:

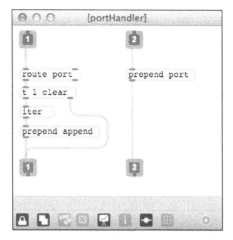

The [p portHandler] subpatcher

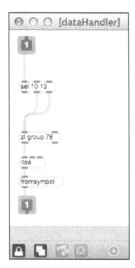

The [p dataHandler] subpatcher

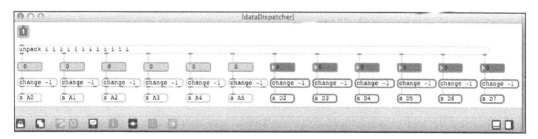

The [p dataDispatcher] subpatcher

In the last figure, we can see only one inlet and no outlets. Indeed, we just encapsulated the global data dispatcher system inside the subpatcher. And this latter generates its data buses with [send] objects. This is an example where we don't need and even don't want to use outlets. Using outlets would be messy because we would have to link each element requesting this or that value from Arduino with a lot of cords.

In order to create a subpatcher, you only have to type n to create a new object, and type p, a space, and the name of your subpatcher.

While I designed these examples, I used something that works faster than creating a subpatcher, copying and pasting the structure on the inside, removing the structure from the outside, and adding inlets and outlets.

This feature is named encapsulate and is part of the **Edit** menu of Max 6.

You have to select the part of the patch you want to encapsulate inside a subpatcher, then click on **Encapsulate**, and voilà! You have just created a subpatcher including your structures that are connected to inlets and outlets in the correct order.

Encapsulate and de-encapsulate features

You can also de-encapsulate a subpatcher. It would follow the opposite process of removing the subpatcher and popping out the whole structure that was inside directly outside.

Subpatching helps to keep things well organized and readable.

We can imagine that we have to design a whole patch with a lot of wizardry and tricks inside it. This one is a processing unit, and as soon as we know what it does, after having finished it, *we don't want to know how it does it* but only *use it*.

This provides a nice abstraction level by keeping some processing units closed inside boxes and not messing the main patch.

You can copy and paste the subpatchers. This is a powerful way to quickly duplicate process units if you need to. But each subpatcher is totally independent of the others. This means that if you need to modify one because you want to update it, you'd have to do that individually in each subpatcher of your patch.

This can be really hard.

Let me introduce you to the last pure Max 6 concept now named **abstractions** before I go further with Arduino.

Abstractions and reusability

Any patch created and saved can be used as a new object in another patch. We can do this by creating a new object by typing n in a patcher; then we just have to type the name of our previously created and saved patch.

A patch used in this way is called an **abstraction**.

In order to call a patch as an abstraction in a patcher, the patch has to be in the Max 6 *path* in order to be found by it. You can check the path known by Max 6 by going to **Options | File Preferences**. Usually, if you put the main patch in a folder and the other patches you want to use as abstractions in that same folder, Max 6 finds them.

The concept of abstraction in Max 6 itself is very powerful because it provides **reusability**.

Indeed, imagine you need and have a lot of small (or big) patch structures that you are using every day, every time, and in almost every project. You can put them into a specific folder on your disk included in your Max 6 path and then you can call (we say *instantiate*) them in every patch you are designing.

Since each patch using it has only a reference to the one patch that was instantiated itself, you just need to improve your abstraction; each time you load a patch using it, the patch will have up-to-date abstractions loaded inside it.

It is really easy to maintain all through the development months or years.

Of course, if you totally change the abstraction to fit with a dedicated project/patch, you'll have some problems using it with other patches. You have to be careful to maintain even short documentation of your abstractions.

Let's now continue by describing some good examples with Arduino.

Creating a sound-level meter with LEDs

This small project is a typical example of a Max 6/Arduino hardware and software collaboration.

Max can easily listen for sounds and convert them from the analog to the digital domain.

We are going to build a small sound level visualizer using Arduino, some LEDs, and Max 6.

The circuit

We are going to use the same circuit we designed in *Chapter 8, Designing Visual Output Feedbacks*, while we multiplexed LEDs with a daisy chain of shift registers of the type 595.

The following figure shows the circuit:

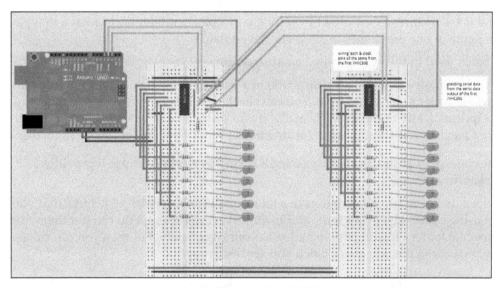

Our double series of eight LEDs

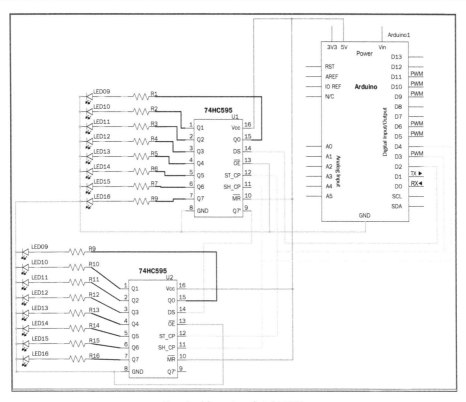

Our double series of eight LEDs

The basic idea is to:

- Use each series of eight LEDs for each sound channel (left and right)
- Display the sound level all along the LED series

For each channel, the greater the number of LEDs switched on, the higher the sound level.

Let's now check how we can handle this in Max 6 first.

The Max 6 patch for calculating sound levels

Have a look at the following figure showing the SoundLevelMeters patch:

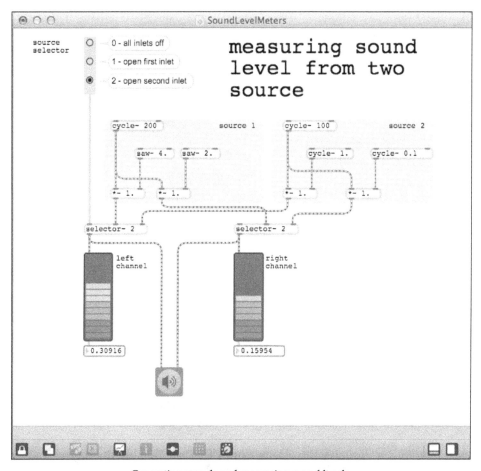

Generating sounds and measuring sound levels

We are using the MSP part of the Max 6 framework here that is related to sound signals. We have two sources (named source 1 and source 2) in the patch. Each one generates two signals. I connected each one to one of the [selector~] objects.

Those latter are switches for signals. The source selector at the top left provides a way to switch between source 1 and source 2.

I won't describe the cheap wizardry of sound sources; it would involve having a knowledge of synthesis and that would be out of the scope of this topic.

Then, we have a connection between each [selector~] output and a small symbol like a speaker. This is related to the sound output of your audio interface.

I also used the [meter~] object to display the level of each channel.

At last, I added a [flonum] object to display the current value of the level each time.

These are the numbers we are going to send to Arduino.

Let's add the serial communication building blocks we already described.

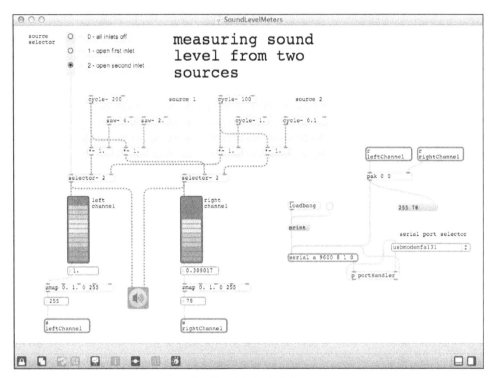

Sending data to Arduino

We have our serial communication setup ready.

We also have the [zmap 0. 1. 0 255] objects. These take a value intended to be between 0. 1, as was set up in the arguments, and scale it to the range 0 255. This provides a byte of data for each channel.

We are using two data buses to send a value from each channel to a [pak] object. The latter collects the incoming messages and creates a list with them. The difference between [pack] and [pak] is that [pak] sends data as soon as it receives a message in one of its inputs, not only when it receives a message of its left input, as with [pack].

Thus, we have lists of messages that are popped out from the computer to Arduino as soon as the level values change.

The firmware for reading bytes

Let's see how to handle this in Arduino:

```
#include <ShiftOutX.h>
#include <ShiftPinNo.h>

int CLOCK_595 = 4;     // first 595 clock pin connecting to pin 4
int LATCH_595 = 3;     // first 595 latch pin connecting to pin 3
int DATA_595 = 2;      // first 595 serial data input pin connecting to
pin 2

int SR_Number = 2;     // number of shift registers in the chain

// instantiate and enabling the shiftOutX library with our circuit
parameters
shiftOutX regGroupOne(LATCH_595, DATA_595, CLOCK_595, MSBFIRST, SR_
Number);

// random groove machine variables
int counter = 0;
byte LeftChannel = B00000000 ;  // store left channel Leds infos
byte RightChannel = B00000000 ; // store right channel Leds infos

void setup() {
  // NO MORE setup for each digital pin of the Arduino
  // EVERYTHING is made by the library :-)
}

void loop(){

  if (Serial.available() > 0) {
    LeftChannel = (byte)Serial.parseInt();
    RightChannel = (byte)Serial.parseInt();


    unsigned short int data; // declaring the data container as a very
local variable
    data = ( LeftChannel << 8 ) | RightChannel; // aggregating the 2
read bytes
    shiftOut_16(DATA_595, CLOCK_595, MSBFIRST, data);  // pushing the
whole data to SRs

    // make a short pause before changing LEDs states
    delay(2);
  }
}
```

This is the same firmware as the one in *Chapter 8, Designing Visual Output Feedback*, except here we are pitreading real values and not generating random ones.

We are doing that with `Serial.parseInt()` in the `Serial.available()` test.

This means that as soon as the data is in the Arduino serial buffer, we'll read it. Actually, we are reading two values and storing them, after a byte conversion, in `LeftChannel` and `RightChannel`.

We then process the data to the shift register to light the LEDs according to the value sent by the Max 6 patch.

Let's take another example of playing with sound files and a distance sensor.

Pitch shift effect controlled by hand

Pitch shifting is a well-known effect in all fields related to sound processing. It changes the pitch of an incoming sound. Here we are going to implement a very cheap pitch shifter with Max 6, but we will focus on how to control this sound effect. We will control it by moving our hand over a distance sensor.

We are going to use the same circuit as the one in *Chapter 6, Sense the World – Feeling with Analog Inputs*.

The circuit with the sensor and the firmware

The following circuit shows the Arduino board connected to a sensor:

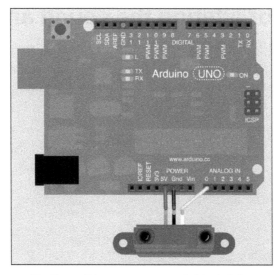

The Sharp distance sensor connected to Arduino

The firmware is almost the same too. I removed the part about the distance calculation because, indeed, we don't care about the distance itself.

The ADC of Arduino provides a resolution of 10 bits, which will give numbers from 0 to 1023. We are going to use this value to calibrate our system.

The following code is the firmware. You can find it in the `Chapter12/PitchShift` folder:

```
int sensorPin = 0;              // pin number where the SHARP GP2Y0A02YK
is connected
int sensorValue = 0 ;           // storing the value measured from 0 to
1023

void setup() {
  Serial.begin(9600);
}

void loop(){
  sensorValue = analogRead(sensorPin); // read/store the value from
sensor
  Serial.println(sensorValue);

  delay(20);
}
```

As soon as Arduino runs this firmware, it sends values to the serial port.

The patch for altering the sound and parsing Arduino messages

I cannot describe the whole pitch shifter itself. By the way, you can open the related subpatch to see how it has been designed. Everything is open.

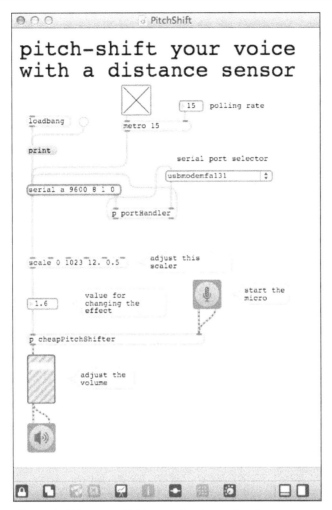

The pitch shifter controlled by your hand over the distance sensor

As we described before, we have to choose the right serial port and then bang the [serial] object in order to make it pop out the values in its buffer.

Here, we are using the [scale] object. It is similar to [zmap], which we already used, because it maps a range to another one but it can also work with inverted range and doesn't clip values.

Here, I'm mapping values being received from the ADC of Arduino from 0 to 1023 to something fitting our need from 12.0 to 0.5.

If we place our hand close to the sensor, the distance is small, and if we move our hand further away, the distance changes and the effect is modulated.

Summary

This chapter taught us how to deal with Arduino using Max 6.

We learnt a bit more about some usual techniques in Max 6, and we practiced some concepts previously learnt in this book. Obviously, there is more to learn in Max 6, and I'd like to give you some good pointers for better learning.

Firstly, I'd suggest you read *all* the tutorials, beginning with those about Max, then about MSP, and then about digital sound, and at last about Jitter if you are interested in visuals and OpenGL. That sounds obvious but I still have two or three persons a day asking me where to begin Max 6 from. The answer is: tutorials.

Then, I'd suggest you design a small system. Less is definitely more. A small system provides easy ways to maintain, modify, and support. Using comments is also a nice way to quickly remember what you tried to do in this or that part.

Lastly, patching a bit everyday is the real key to success. It takes time, but don't we want to become masters?

13
Improving your C Programming and Creating Libraries

This is the last chapter of this book and is the most advanced, but not the most complex. You will learn about C code optimization through several typical examples that will bring you a bit further and make you more capable for your future projects using Arduino. I am going to talk about libraries and how they can improve the reusability of your code to save time in the future. I will describe some tips to improve the performance of your code by using bit-shifting instead of the usual operators, and by using some memory management techniques. Then, I will talk about reprogramming the Arduino chip itself and debugging our code using an external hardware programmer.

Let's go.

Programming libraries

I have already spoken about libraries in *Chapter 2, First Contact with C*. We can define it as a set of implementations of behavior already written using a particular language that provides some interfaces by which all the available behaviors can be called.

Basically, a library is something already written and reusable in our own code by following some specifications. For example, we can quote some libraries included in the Arduino core. Historically, some of those libraries had been written independently, and over time, the Arduino team as well as the whole Arduino community incorporated them into the growing core as natively available libraries.

Let's take the EEPROM library. In order to check files related to it, we have to find the right folder on our computer. On OS X, for instance, we can browse the contents of the `Arduino.app` file itself. We can go to the `EEPROM` folder in `Contents/Resources/Java/libraries/`. In this folder, we have three files and a folder named `examples` containing all the examples related to the EEPROM library:

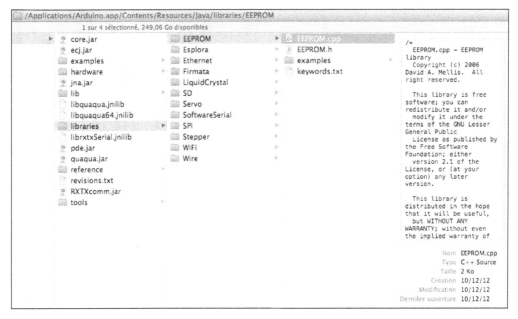

The EEPROM library on our computer (an OS X system)

We have the following files:

- `EEPROM.h`, containing the headers of the library
- `EEPROM.cpp`, containing the code itself
- `keywords.txt`, containing some parameters to color the keywords of the library

Because of the location of these files in the folder hierarchy, they are available as parts of the core EEPROM library. This means we can include this library as soon as we have the Arduino environment installed on our computer without downloading anything else.

The simple statement `include <EEPROM.h>` includes the library in our code and makes all the features of this library available for further use.

Let's enter code in these files.

The header file

Let's open EEPROM.h:

```
/*
  EEPROM.h - EEPROM library
  Copyright (c) 2006 David A. Mellis.  All right reserved.

  This library is free software; you can redistribute it and/or
  modify it under the terms of the GNU Lesser General Public
  License as published by the Free Software Foundation; either
  version 2.1 of the License, or (at your option) any later version.

  This library is distributed in the hope that it will be useful,
  but WITHOUT ANY WARRANTY; without even the implied warranty of
  MERCHANTABILITY or FITNESS FOR A PARTICULAR PURPOSE.  See the GNU
  Lesser General Public License for more details.

  You should have received a copy of the GNU Lesser General Public
  License along with this library; if not, write to the Free Software
  Foundation, Inc., 51 Franklin St, Fifth Floor, Boston, MA  02110-1301  USA
*/

#ifndef EEPROM_h
#define EEPROM_h

#include <inttypes.h>

class EEPROMClass
{
  public:
    uint8_t read(int);
    void write(int, uint8_t);
};

extern EEPROMClass EEPROM;

#endif
```

EEPROM.h displayed in Xcode IDE

In this file, we can see some preprocessor directives starting with the # character. This is the same one that we use to include libraries in our Arduino code. Here, this is a nice way to not include the same header twice. Sometimes, while coding, we include a lot of libraries and at compilation time, we would have to check that we didn't include the same code twice. These directives and especially the ifndef directive mean: "If the EEPROM_h constant has not been defined, then do the following statements".

This is a trick commonly known as **include guards**. The first thing we are doing after this test is defining the EEPROM_h constant. If in our code we or some other libraries include the EEPROM library, the preprocessor wouldn't reprocess the following statements the second time it sees this directive.

We have to finish the #ifndef directive with the #endif directive. This is a common block in the header files and you'll see it many times if you open other library header files files. What is contained inside this block? We have another inclusion related to C integer types: #include <inttypes.h>.

The Arduino IDE contains all the required C headers in the library. As we have already mentioned, we could use pure C and C++ code in our firmware. We didn't until now because the functions and types we've been using have already been coded into the Arduino core. But please keep in mind that you have the choice to include other pure C code in your firmware and in this last chapter, we will also talk about the fact you can also follow pure AVR processor-type code too.

Now we have a class definition. This is a C++ feature. Inside this class, we declare two function prototypes:

- uint8_t read(int)
- void write(int, uint8_t)

There is a function to read something, taking an integer as an argument and returning an unsigned integer that is 8 bits long (which is a byte). Then, there is another function to write something that takes an integer and a byte and returns nothing. These prototypes refer to the definition of these functions in the other EEPROM.cpp file.

The source file

Let's open EEPROM.cpp:

The source file of the EEPROM library is displayed in the Xcode IDE

The file begins by including some headers. `avr/eeprom.h` refers to the AVR type processor's EEPROM library itself. In this library example, we just have a library referring to and making a better interface for our Arduino programming style than the original pure AVR code. This is why I chose this library example. This is the shortest but the most explicit example, and it teaches us a lot.

Then we include the `Arduino.h` header in order to have access to standard types and constants of the Arduino language itself. At last, of course, we include the header of the EEPROM library itself.

In the following statements, we define both functions. They call other functions inside their block definition:

- `eeprom_read_byte()`
- `eeprom_write_byte()`

Those functions come from the AVR EEPROM library itself. The EEPROM Arduino library is only an interface to the AVR EEPROM library itself. Why wouldn't we try to create a library ourselves?

Creating your own LED-array library

We are going to create a very small library and test it with a basic circuit including six LEDs that are not multiplexed.

Wiring six LEDs to the board

Here is the circuit. It basically contains six LEDs wired to Arduino:

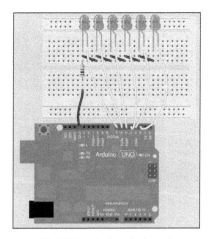

Six LEDs wired to the board

The circuit diagram is shown as follows:

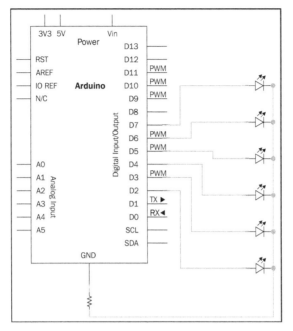

Another diagram of the six LEDs wired directly to Arduino

I won't discuss the circuit itself, except to mention that I put in a 1 kΩ resistor. I took the worst case where all LEDs would be switched on at the same time. This would drive a lot of current, and so this acts as security for our Arduino. Some authors wouldn't use it. I'd prefer to have some LEDs dimming a bit in order to protect my Arduino.

Creating some nice light patterns

Here is code for lighting up the LEDs according to some patterns, all hardcoded. A pause is made between each pattern display:

```
void setup() {

  tor (int i = 2 ; i <= 7 ; i++)
  {
    pinMode(i, OUTPUT);
  }
}

void loop(){

  // switch on everything progressively
```

```
for (int i = 2 ; i <= 7 ; i++)
{
  digitalWrite(i, HIGH);
  delay(100);
}

delay(3000);

// switch off everything progressively
for (int i = 7 ; i >=2 ; i--)
{
  digitalWrite(i, LOW);
  delay(100);
}

delay(3000);

// switch on even LEDS
for (int i = 2 ; i <= 7 ; i++)
{
  if ( i % 2 == 0 ) digitalWrite(i, HIGH);
  else digitalWrite(i, LOW);
}

delay(3000);

// switch on odd LEDS
for (int i = 2 ; i <= 7 ; i++)
{
  if ( i % 2 != 0 ) digitalWrite(i, HIGH);
  else digitalWrite(i, LOW);
}

delay(3000);
}
```

This code works correctly. But how could we make it more elegant and, especially, more reusable? We could embed the for() blocks into functions. But these would only be available in this code. We'd have to copy and paste them by remembering the project in which we designed them in order to reuse them in another project.

By creating a small library that we can use over and over again, we can save time in the future in coding as well as processing. With some periodic modifications, we can arrive at the perfect module for its intended task, which will get better and better until there's no need to even touch it anymore because it performs more perfectly than anything else out there. At least that's what we hope for.

Designing a small LED-pattern library

At first, we can design our function's prototype in a header. Let's call the library
`LEDpatterns`.

Writing the LEDpatterns.h header

Here is how a possible header could be:

```
/*
  LEDpatterns - Library for making cute LEDs Pattern.
  Created by Julien Bayle, February 10, 2013.
*/
#ifndef LEDpatterns_h
#define LEDpatterns_h

#include "Arduino.h"

class LEDpatterns
{
  public:
    LEDpatterns(int firstPin, int ledsNumber);
    void switchOnAll();
    void switchOffAll();
    void switchEven();
    void switchOdd();
  private:
    int _firstPin;
    int _ledsNumber;
};
#endif
```

We first write our include guards. Then we include the Arduino library. Then, we
define a class named `LEDpatterns` with the `public` functions including a constructor
that has the same name as the class itself.

We also have two internal (`private`) variables related to the first pin on which LEDs
are wired and related to the total number of LEDs wired. LEDs would have to be
contiguously wired in that example.

Writing the LEDpatterns.cpp source

Here is the source code of the C++ library:

```cpp
/*
  LEDpatterns.cpp - Library for making cute LEDs Pattern.
  Created by Julien Bayle, February 10, 2013.
  */
#include "Arduino.h"
#include "LEDpatterns.h"

LEDpatterns::LEDpatterns(int firstPin, int ledsNumber)
{
  for (int i = firstPin ; i < ledsNumber + firstPin ; i++)
  {
    pinMode(i, OUTPUT);
  }

  _ledsNumber = ledsNumber;
  _firstPin = firstPin;
}

void LEDpatterns::switchOnAll()
{
  for (int i = _firstPin ; i < _ledsNumber + _firstPin ; i++)
  {
    digitalWrite(i, HIGH);
    delay(100);
  }
}

void LEDpatterns::switchOffAll()
{
  for (int i = _ledsNumber + _firstPin -1 ; i >= _firstPin    ; i--)
  {
    digitalWrite(i, LOW);
    delay(100);
  }
}

void LEDpatterns::switchEven()
{
  for (int i = _firstPin ; i < _ledsNumber + _firstPin ; i++)
  {
    if ( i % 2 == 0 ) digitalWrite(i, HIGH);
    else digitalWrite(i, LOW);
  }
}

void LEDpatterns::switchOdd()
```

```
{
    for (int i = _firstPin ; i < _ledsNumber + _firstPin ; i++)
    {
        if ( i % 2 != 0 ) digitalWrite(i, HIGH);
        else digitalWrite(i, LOW);
    }
}
```

At the beginning, we retrieve all the include libraries. Then we have the constructor, which is a special method with the same name as the library. This is the important point here. It takes two arguments. Inside its body, we put all the pins from the first one to the last one considering the LED number as a digital output. Then, we store the arguments of the constructor inside the private variables previously defined in the header LEDpatterns.h.

We can then declare all our functions related to those created in the first example without the library. Notice the LEDpatterns:: prefix for each function. I won't discuss this pure class-related syntax here, but keep in mind the structure.

Writing the keyword.txt file

When we look at our source code, it's very helpful to have things jump out at you, and not blend into the background. In order to correctly color the different keywords related to our new created library, we have to use the keyword.txt file. Let's check this file out:

```
#######################################
# Syntax Coloring Map For Messenger
#######################################


#######################################
# Datatypes (KEYWORD1)
#######################################


LEDpatterns          KEYWORD1


#######################################
# Methods and Functions (KEYWORD2)
#######################################
switchOnAll          KEYWORD2
switchOffAll         KEYWORD2
switchEven           KEYWORD2
switchOdd KEYWORD2


#######################################
```

```
# Instances (KEYWORD2)
#####################################


#####################################
# Constants (LITERAL1)
#####################################
```

In the preceding code we can see the following:

- Everything followed by KEYWORD1 will be colored in orange and is usually for classes
- Everything followed by KEYWORD2 will be colored in brown and is for functions
- Everything followed by LITERAL1 will be colored in blue and is for constants

It is very useful to use these in order to color your code and make it more readable.

Using the LEDpatterns library

The library is in the LEDpatterns folder in Chapter13 and you have to put it in the correct folder with the other libraries, which we have done. We have to restart the Arduino IDE in order to make the library available. After having done that, you should be able to check if it is in the menu **Sketch | Import Library**. LEDpatterns is now present in the list:

The library is a contributed one because it is not part of the Arduino core

Let's now check the new code using this library. You can find it in the Chapter13/ LEDLib folder:

```
#include <LEDpatterns.h>
LEDpatterns ledpattern(2,6);

void setup() {
}

void loop(){

  ledpattern.switchOnAll();
  delay(3000);

  ledpattern.switchOffAll();
  delay(3000);

  ledpattern.switchEven();
  delay(3000);

  ledpattern.switchOdd();
  delay(3000);
}
```

In the first step, we include the LEDpatterns library. Then, we create the instance of LEDpatterns named ledpattern. We call the constructor that we designed previously with two arguments:

- The first pin of the first LED
- The total number of LEDs

ledpattern is an instance of the LEDpatterns class. It is referenced throughout our code, and without #include, it would not work. We have also invoked each method of this instance.

If the code seems to be cleaner, the real benefit of such a design is the fact that we can reuse this library inside any of our projects. If we want to modify and improve the library, we only have to modify things in the header and the source file of our library.

Memory management

This section is a very short one but not a less important one at all. We have to remember we have the following three pools of memory on Arduino:

- Flash memory (program space), where the firmware is stored
- **Static Random Access Memory** (**SRAM**), where the sketch creates and manipulates variables at runtime
- EEPROM is a memory space to store long-term information

Flash and EEPROM, compared to SRAM, are non-volatile, which means the data persists even after the power is turned off. Each different Arduino board has a different amount of memory:

- ATMega328 (UNO) has:
  - Flash 32k bytes (0.5k bytes used by the bootloader)
  - SRAM 2k bytes
  - EEPROM 1k bytes
- ATMega2560 (MEGA) has:
  - Flash 256k bytes (8k bytes used by the bootloader)
  - SRAM 8k bytes
  - EEPROM 4k bytes

A classic example is to quote a basic declaration of a string:

```
char text[] = "I love Arduino because it rocks.";
```

That takes 32 bytes into SRAM. It doesn't seem a lot but with the UNO, you *only* have 2048 bytes available. Imagine you use a big lookup table or a large amount of text. Here are some tips to save memory:

- If your project uses both Arduino and a computer, you can try to move some calculation steps from Arduino to the computer itself, making Arduino only trigger calculations on the computer and request results, for instance.
- Always use the smallest data type possible to store values you need. If you need to store something between 0 and 255, for instance, don't use an int type that takes 2 bytes, but use a byte type instead
- If you use some lookup tables or data that won't be changed, you can store them in the Flash memory instead of the SRAM. You have to use the PROGMEM keyword to do that.

- You can use the native EEPROM of your Arduino board, which would require making two small programs: the first to store that information in the EEPROM, and the second to use it. We did that using the PCM library in the *Chapter 9, Making Things Move and Creating Sounds*.

Mastering bit shifting

There are two bit shift operators in C++:

- `<<` is the left shift operator
- `>>` is the right shift operator

These can be very useful especially in SRAM memory, and can often optimize your code. `<<` can be understood as a multiplication of the left operand by 2 raised to the right operand power.

`>>` is the same but is similar to a division. The ability to manipulate bits is often very useful and can make your code faster in many situations.

Multiplying/dividing by multiples of 2

Let's multiply a variable using bit shifting.

```
int a = 4;
int b = a << 3;
```

The second row multiplies the variable a by 2 to the third power, so b now contains 32. On the same lines, division can be carried out as follows:

```
int a = 12 ;
int b = a >> 2;
```

b contains 3 because `>> 2` equals division by 4. The code can be faster using these operators because they are a direct access to binary operations without using any function of the Arduino core like `pow()` or even the other operators.

Packing multiple data items into bytes

Instead of using a big, two-dimensional table to store, for instance, a bitmap shown as follows:

```
const prog_uint8_t BitMap[5][7] = {
// store in program memory to save RAM
{1,1,0,0,0,1,1},
```

```
{0,0,1,0,1,0,0},
{0,0,0,1,0,0,0},
{0,0,1,0,1,0,0},
{1,1,0,0,0,1,1}      };
```

We can use use the following code:

```
const prog_uint8_t BitMap[5] = {
// store in program memory to save RAM
B1100011,
B0010100,
B0001000,
B0010100,
B1100011     };
```

In the first case, it takes 7 x 5 = 35 bytes per bitmap. In the second one, it takes only 5 bytes. I guess you've just figured out something huge, haven't you?

Turning on/off individual bits in a control and port register

The following is a direct consequence of the previous tip. If we want to set up pins 8 to 13 as output, we could do it like this:

```
void setup()      {
  int pin;

  for (pin=8; pin <= 13; ++pin) {
    pinMode (pin, LOW);
  }
}
```

But this would be better:

```
void setup()      {
 DDRB = B00111111 ; // DDRB are pins from 8 to 15
}
```

In one pass, we've configured the whole package into one variable directly in memory, and no `pinMode` function, structure, or variable name needs to be compiled.

Reprogramming the Arduino board

Arduino natively uses the famous bootloader. This provides a nice way to upload our firmware using the virtual serial port on the USB. But we might be interested to go ahead without any bootloader. How and why? Firstly, that would save some Flash memory. It also provides a way to avoid the small delay when we power on or reset our board before it becomes active and starts running. It requires an external programmer.

I can quote the AVR-ISP, the STK500, or even a parallel programmer (a parallel programmer is described at `http://arduino.cc/en/Hacking/ ParallelProgrammer`). You can find an AVR-ISP at Sparkfun Electronics.

I used this one a couple of times to program an Arduino FIO-type board for specific wireless applications in a project connecting cities named The Village in 2013.

The Pocket AVR programmer by Sparkfun Electronics

This programmer can be wired using 2 x 5 connectors to the ICSP port on the Arduino board.

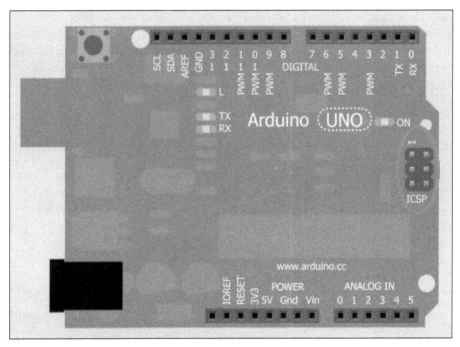

The ICSP connector of Arduino

In order to reprogram the processor of Arduino, we have to first close the Arduino IDE, and then check the preferences file (`preferences.txt` on a Mac, located in `Contents/Resources/Java/lib` inside the `Arduino.app` package itself). On a Windows 7 PC and higher, this file is located at: `c:\Users\<USERNAME>\AppData\Local\Arduino\preferences.txt`. In Linux it is located at: `~/arduino/preferences.ard`.

We have to change the `upload.using` value that is initially set to bootloader to the correct identifier that fits your programmer. This can be found in the content of the Arduino application package on OS X or inside the Arduino folders on Windows. For instance, if you display the `Arduino.app` content, you can find this file: `Arduino.app/Contents/Resources/Java/hardware/arduino/programmers.txt`.

Then we can start the Arduino IDE to upload the sketch using our programmer. To revert back to the normal bootloader behavior, we have to first reupload the bootloader that fits with our hardware. Then, we have to change back the `preferences.txt` file, and it will work as the initial board.

Summary

In this chapter, we learned more about designing libraries, and we are now able to design our projects a bit differently, keeping in mind reusability of the code or part of the code in future projects. This can save time and also improves readability.

We can also explore existing libraries and enjoy the world of open source by taking them, hacking them, and making them fit our needs. This is a really open world into which we have just made our first steps.

Conclusion

We are at the end of this book. You have probably read everything and also tested some pieces of code with your own hardware, and I'm sure you are now able to imagine your future and advanced projects with Arduino.

I wanted to thank you for being so focused and interested. I know you are now almost in the same boat as myself, you want to learn more, test more, and check and use new technologies in order to achieve your craziest project. I'd like to say one last thing: do it, and do it now!

In most cases, people are afraid of the huge amount of work that they can imagine in the first steps just before they start. But you have to trust me, don't think too much about details or about optimization. Try to make something simple, something that works. Then you'll have ways to optimize and improve it.

One last piece of advice for you: don't think too much, and make a lot. I have seen too many unfinished projects by people having wanted to think, think, think instead of just starting and making.

Take care and continue exploring!

Index

Thank you for buying
C Programming for Arduino

About Packt Publishing

Packt, pronounced 'packed', published its first book "*Mastering phpMyAdmin for Effective MySQL Management*" in April 2004 and subsequently continued to specialize in publishing highly focused books on specific technologies and solutions.

Our books and publications share the experiences of your fellow IT professionals in adapting and customizing today's systems, applications, and frameworks. Our solution based books give you the knowledge and power to customize the software and technologies you're using to get the job done. Packt books are more specific and less general than the IT books you have seen in the past. Our unique business model allows us to bring you more focused information, giving you more of what you need to know, and less of what you don't.

Packt is a modern, yet unique publishing company, which focuses on producing quality, cutting-edge books for communities of developers, administrators, and newbies alike. For more information, please visit our website: www.packtpub.com.

About Packt Open Source

In 2010, Packt launched two new brands, Packt Open Source and Packt Enterprise, in order to continue its focus on specialization. This book is part of the Packt Open Source brand, home to books published on software built around Open Source licences, and offering information to anybody from advanced developers to budding web designers. The Open Source brand also runs Packt's Open Source Royalty Scheme, by which Packt gives a royalty to each Open Source project about whose software a book is sold.

Writing for Packt

We welcome all inquiries from people who are interested in authoring. Book proposals should be sent to author@packtpub.com. If your book idea is still at an early stage and you would like to discuss it first before writing a formal book proposal, contact us; one of our commissioning editors will get in touch with you.

We're not just looking for published authors; if you have strong technical skills but no writing experience, our experienced editors can help you develop a writing career, or simply get some additional reward for your expertise.

Raspberry Pi Media Center

ISBN: 978-1-782163-02-2   Paperback: 108 pages

Transform your Raspberry into a full-blown media center within 24 hours

1. Discover how you can stream video, music, and photos straight to your TV

2. Play existing content from your computer or USB drive

3. Watch and record TV via satellite, cable, or terrestrial

4. Build your very own library that automatically includes detailed information and cover material

Raspberry Pi Home Automation with Arduino

ISBN: 978-1-849695-86-2   Paperback: 380 pages

Automate your home with a set of exciting projects for the Raspberry Pi!

1. Learn how to dynamically adjust your living environment with detailed step-by-step examples

2. Discover how you can utilize the combined power of the Raspberry Pi and Arduino for your own projects

3. Revolutionize the way you interact with your home on a daily basis

Please check **www.PacktPub.com** for information on our titles

Raspberry Pi Networking Cookbook

ISBN: 978-1-849694-60-5   Paperback: 204 pages

An epic collection of practical and engaging recipies for the Raspberry Pi!

1. Learn how to install, administer, and maintain your Raspberry Pi

2. Create a network fileserver for sharing documents, music, and videos

3. Host a web portal, collaboration wiki, or even your own wireless access point

4. Connect to your desktop remotely, with minimum hassle

Processing 2: Creative Programming Cookbook

ISBN: 978-1-849517-94-2   Paperback: 436 pages

Over 90 highly-effective recipies to unleash your creativity with interactive art, graphics, computer vision, 3D and more

1. Explore the Processing language with a broad range of practical recipes for computational art and graphics

2. Wide coverage of topics including interactive art, computer vision, visualization, drawing in 3D, and much more with Processing

3. Create interactive art installations and learn to export your artwork for print, screen, Internet, and mobile devices

Please check **www.PacktPub.com** for information on our titles

www.ingramcontent.com/pod-product-compliance
Lightning Source LLC
Chambersburg PA
CBHW080133060326
40689CB00018B/3775